AF364430

Pharmacognosy and Phytochemistry-II

Theory and Practical

Pharmacognosy and Phytochemistry-II

Theory and Practical

Sharada L. Deore

M.Pharm (Pharmacognosy & Phytochemistry),
Ph.D (Pharmaceutical Sciences), PG Diploma in Patent's law
Associate Professor,
Govt. College of Pharmacy, Amravati-444604
Maharashtra.

PharmaMed Press

An imprint of BSP Books Pvt. Ltd.
4-4-309/316, Giriraj Lane,
Sultan Bazar, Hyderabad - 500 095.

Pharmacognosy and Phytochemistry-II, Theory and Practical

by Sharada L. Deore

> ***Disclaimer:*** The authors and the publishers have taken due care to provide the authentic, reliable and up to date information related to the subject. However, neither the authors nor the publisher shall be responsible for any liability for any damage caused as a result of use of this book. The respective user must check the accuracy from other sources too.

Published by

PharmaMed Press

An imprint of BSP Books Pvt. Ltd.

4-4-309/316, Giriraj Lane, Sultan Bazar, Hyderabad - 500 095.

Phone: 040-23445688; Fax: 91+40-23445611

E-mail: info@pharmamedpress.net

www.bspbooks.net/www.pharmamedpress.net

ISBN: 978-93-91910-71-6 (Hardback)

Preface

The science of pharmacognosy dealt mostly with physical description and identification of whole and powdered plant drugs including their history, commerce, collection, preparation, and storage. Since 20th century, remarkable progress in biological and chemical techniques facilitated the understanding of biogenesis, isolation and structure determination of phytochemicals. Role of sophisticated analytical techniques in discovery of novel phytochemicals is noteworthy.

Pharmacy Council of India (PCI) has introduced ***Pharmacognosy and Phytochemistry-*II *in B.Pharm Fifth Semester* to make students know and comprehend about biogenesis of primary and secondary metabolites, isolation, identification and estimation of phytochemicals and applications of important analytical techniques.

Experimental skills of qualitative as well as quantitative phytochemical screening, isolation of therapeutically important phytochemicals, chromatographic analysis of phytochemicals will help student to serve better in community, herbal industry, academics and research establishments.

Present book is strictly designed as per PCI syllabus to cover theory as well as practical syllabus topics to able students to upgrade the desired theoretical together with practical skills.

Subjective as well as specific and related objective (MCQs) questions are added to analyse the topic understanding of students.

I would first and foremost like to acknowledge the authors and publishers of various books, research articles, journals, websites and other sources that have been referred to for putting together to make this book a fine source of reference.

Author is thankful to all her teachers, students, friends and family for motivation of keeping writing as per the need of time and changing scenario.

Author would like to acknowledge the excellent efforts of Publisher Anil Shah and Editor Naresh Daver for corrections and suggestions to bring out the theory as well as practical syllabus content in the form of semester book.

Sharada Deore-Baviskar

Contents

UNIT 3

Isolation, Identification and Analysis of Phytoconstituents

UNIT 4

Industrial Production, Estimation and Utilization

UNIT 5

Basics of Phytochemistry

Part – II

Practical Manual

Part – I
(Theory)

Pharmacognosy and Phytochemistry-II

Unit1

Metabolic Pathways in Higher Plants and their Determination

PCI Syllabus

Metabolic pathways in higher plants and their determination

- Brief study of basic metabolic pathways and formation of different secondary metabolites through these pathways- Shikimic acid pathway, Acetate pathways and Amino acid pathway.

- Study of utilization of radioactive isotopes in the investigation of Biogenetic studies

---

Chapter Content

1.1. Brief Study of basic Metabolic Pathways

    1.1.1   Photosynthesis

    1.1.2   Glycolysis

    1.1.3   Citric Acid Cycle

    1.1.4   Pentose Phosphate Pathway

1.2. Acetate Pathway

    1.2.1   Introduction

    1.2.2   Saturated Fatty Acid Biosynthesis

    1.2.3   Unsaturated Fatty Acid (UFA) Biosynthesis

1.3. Shikimic Acid Pathways

1.4. Amino Acid Biosynthesis Pathway

1.5. Elucidation of Biosynthetic Pathway

1.1   Brief Study of Basic Metabolic Pathways

Biogenesis or *in-vivo* synthesis of both primary and secondary metabolites starts with photosynthesis to produce sugar molecules which are metabolised to glycerates, pyruvates and finally acetyl CoA. This acetyl CoA is used in TCA cycle to generate a number of amino acids and excess is to synthesize fatty acids. Few of the acetyl CoA molecules are condensed to form mevalonic acid, precursor of synthesis of steroids and terpenoides. Amino acids give rise to alkaloids. The intermediates of glycolysis i.e. glyceraldehyde 3-phosphate and erythrose 4-phopshate from pentose phosphate pathway yields shikimic acid which is main precursor for biosynthesis of number of important aromatic chemicals like phenylpropanoides, lignin, lignans, flavonoides and terpenoid quinones.

1.1.1   Photosynthesis

Photosynthesis means "putting together with light". Photosynthesis in green plants and specialized bacteria is the process of utilizing light energy to synthesize organic compounds from carbon dioxide and water. Plants absorb light primarily using the pigment chlorophyll, which is the reason that most plants have a green colour. Besides chlorophyll, plants also use pigments such as carotenes and xanthophylls. Carbon dioxide and oxygen enter and leave through tiny pores called *stomata*. It consists of the light dependent part (light reaction) and the light independent part (dark reaction, carbon fixation).

The general equation for photosynthesis is therefore:

$$2n\ CO_2 + 2n\ H_2O + Sunlight \rightarrow 2(CH_2O)n + n\ O_2 + 2n\ A$$

Carbon dioxide + Electron donor + Light energy $\rightarrow$ Carbohydrate + Oxygen + Oxidized electron donor

In the first stage, light-dependent reactions capture the energy of light and use it to make the energy-storage molecules ATP and NADPH via ATP synthase.

In the second stage, the light-independent reactions together known as *Calvin cycle*(Fig. 1.1) reduce carbon dioxide via enzyme RuBisCO (Ribulose-1, 5-bisphosphate carboxylase oxygenase). The product of the Calvin cycle is 3-carbon compound glyceraldehyde-3-phosphate and water. Two molecules of glyceraldehyde-3-phosphate combine to form one molecule of glucose and later different larger carbohydrates. The overall equation for the light-dependent reactions is:

$$2\ H_2O + 2\ NADP^+ + 2\ ADP + 2\ Pi + Light \rightarrow 2\ NADPH + 2\ H^+ + 2\ ATP + O_2$$

The overall equation for the light-independent reactions is :

$$3\ CO_2 + 9\ ATP + 6\ NADPH + 6\ H^+ \rightarrow C_3H_6O_3\text{-phosphate} + 9\ ADP + 8\ Pi + 6\ NADP^+ + 3H_2O$$

In light independent part the carbon fixation refers to any process through which gaseous carbon dioxide is converted into a solid compound like sugar molecules. There are three types of Carbon fixation: C_3, C_4 and CAM. The difference between C_3 and C_4 photosynthesis depends on differences in the chemical compounds to which the incoming CO_2 is linked during the dark reactions (CAM photosynthesis differs from both C_3 and C_4 photosynthesis in that prior to fixation, CO2 is an acid form known as carbonic acid).

Differences between different Photosynthetic Pathways

C_3 pathway	C_4 pathway	CAM pathway
Calvin cycle or C_3 cycle or Calvin–Benson-Bassham cycle or CBB cycle or reductive pentose phosphate cycle is the most well-known type of photosyn-thesis.	Hatch and Slack's C4 photosynthesis.	Crassulacean Acid Metabolism (CAM) is daytime photosynthesis behind closed stomata through pre-fixed CO_2 during night.
First stable product is *a 3-carbon compound* called *phosphoglyceric acid.*	The first stable compound is a *4-C compounds oxalo acetic acid.*	the first stable compound is a *4-C compounds oxalo acetic acid*
Bundle sheath lacking chloroplasts	Mesophyll cells or bundle sheath having chloroplasts	Vacuoles in mesophyll cells
Enzyme Rubisco (*RuBP carboxylase)*	Pepco- enzyme (phosphoenol pyruvate *carboxylase)*	Pepco- enzyme (phosphoenol pyruvate *carboxylase)*
More efficient than C_4 or CAM photosynthesis when the environment is cool and water and light is plentiful.	Photosynthetically more efficient because the net requirement of ATP and $NADPH_2$ for the fixation of one molecule of CO_2 is considerably lower in C_4 plants than in C_3 plants.	Not efficient as C_3 or C_4 but advantage of CAM is that photosynthesis can proceed during the day while the stomata are closed, greatly reducing H_2O loss.
RuBP and CO_2 produces unstable 6-C Compound and gets cleaved to form two molecules of 3C compounds called phosphoglyceric acid (PGA).	During day times, oxaloacetate is converted into malate (malic acid), aspartate and then transported to inner compartment of bundle sheath where deacrboxylated to pyruvates.	During night, CAM plants through open stomata takes CO_2 and fixed it to organic acids stored in vacuoles. During the day CO_2 released to the Calvin cycle to build branched carbohydrates.
Requires more quantity of water hence optimum	Optimum temperature required is 30-40°C.	as stomata remain shut during the photosynthesis

Contd...

temperature required is *cool* (20-25°C)		causes less water loss so useful for plants growing in *hot*, dry day with cool nights or *stressful and arid conditions.*
Example- rice, wheat, potato	Example-maize, sugarcane	Example- pineapple, cacti, orchids

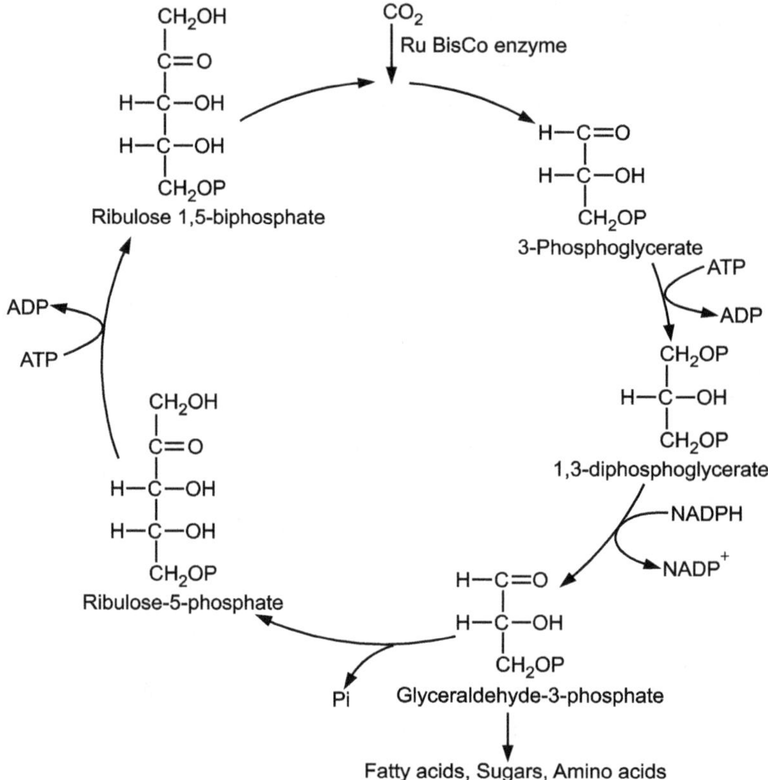

Fig.1.1 Calvin cycle

1.1.2 Glycolysis

Glycolysis (Fig.1.2) is the process of enzymatic reactions that convert glucose into three-carbon compounds, (pyruvate and glycerates) small amounts of ATP (energy) and NADH (reducing power). The glycolytic pathway operates in both situations,in the presence (aerobic) and absence (anaerobic) of oxygen. Under anaerobic conditions, the metabolism of each glucose molecule yields only two ATPs. In contrast, the complete aerobic metabolismof glucose to carbon dioxide by glycolysis and Krebs cycle yields up to thirty-eight ATPs. It is a central pathway that produces important precursor metabolites: six-carbon compounds of glucose-6P and fructose-6P and three-carbon compounds of glycerone-P, glyceraldehyde-3P, glycerate-3P, phosphoenolpyruvate and pyruvate. Acetyl-CoA, another important precursor metabolite, is produced by oxidative decarboxylation of pyruvate in the presence of pyruvate dehydrogenase to form acetyl coenzyme A (acetyl CoA). Under conditions where energy is needed, acetyl CoA is metabolized by Krebs cycle to generate carbon dioxide and a large amount of ATP. When the cell does not need energy, acetyl CoA can be used to synthesize fats or amino acids.

As the glucose is oxidized by the glycolytic enzymes, the coenzyme nicotinamide adenine dinucleotide (NAD^+) is converted from its oxidized to reduced form (NADH). When oxygen is available (aerobic conditions), NADH can reoxidize to NAD^+. However, if either oxygen levels are insufficient (anaerobic conditions) or mitochondrial activity is absent, NADH must be reoxidized by the cell using some other mechanism. In animal cells, the reoxidation of NADH is accomplished by reducing pyruvate, the end-product of glycolysis, to form lactic acid. This process is known as *anaerobic glycolysis*. During vigorous exercise, skeletal muscles relie heavily on it. In yeast, anaerobic conditions result in the production of carbon dioxide and ethanol from pyruvate rather than lactic acid. This process, known as *alcoholic fermentation*, is the basis of wine production and the reason why bread dough rises.

Although some cells are highly dependent on glycolysis for the generation of ATP, the amount of ATP generated per glucose molecule is actually quite small. Therefore, in the majority of cells the most important function of glycolysis is to metabolize glucose to generate three-carbon compounds that can be utilized by other pathways. The final product of aerobic glycolysis is pyruvate.

1.1.3   Citric Acid Cycle

Citric acid cycle also known as the tricarboxylic acid (TCA) cycle or the Krebs cycle (Fig.1.3) is the common mode of oxidative degradation of carbohydrates, fatty acids and amino acids. The cycle starts with acetyl-CoA, the activated form of acetate, derived from glycolysis and pyruvate oxidation of carbohydrates and from beta oxidation of fatty acids. The two-carbon acetyl group in acetyl-CoA is transferred to the four-carbon compound of oxaloacetate to form the six-carbon compound of citrate. In a series of reactions two carbons in citrate are oxidized to CO_2 and the reaction pathway supplies NADH for use in the oxidative phosphorylation and other metabolic processes. The pathway also supplies important precursor metabolites including 2-oxoglutarate. At the end of the cycle the remaining four-carbon part is transformed back to oxaloacetate. Because two acetyl-CoA molecules are produced from each glucose molecule, two cycles are required per glucose molecule. Therefore, at the end of two cycles, the products are: 6 molecules of NADH, two molecules of $FADH_2$, two molecules of ATP, and four molecules of CO_2.

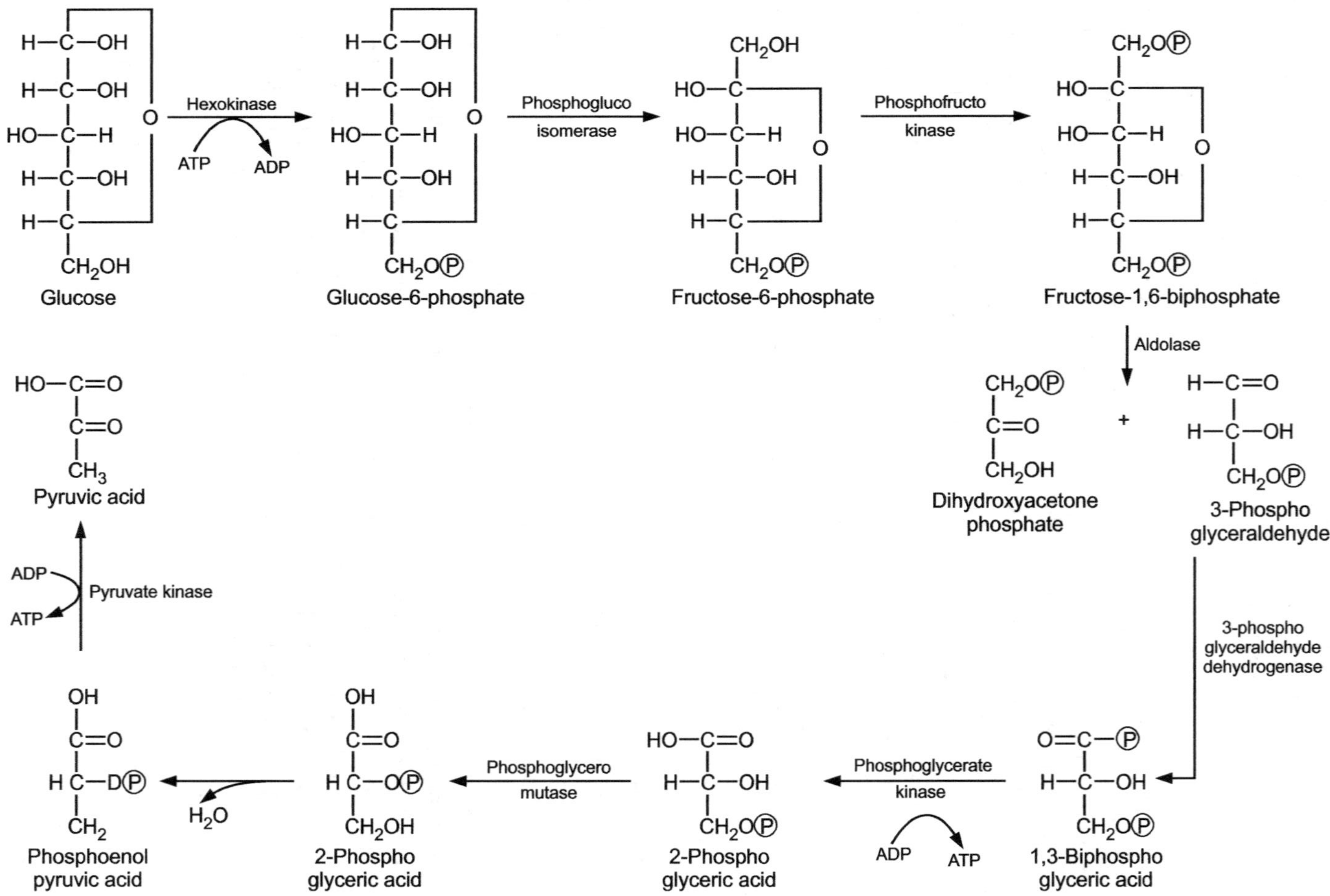

Fig.1.2 Glycolysis

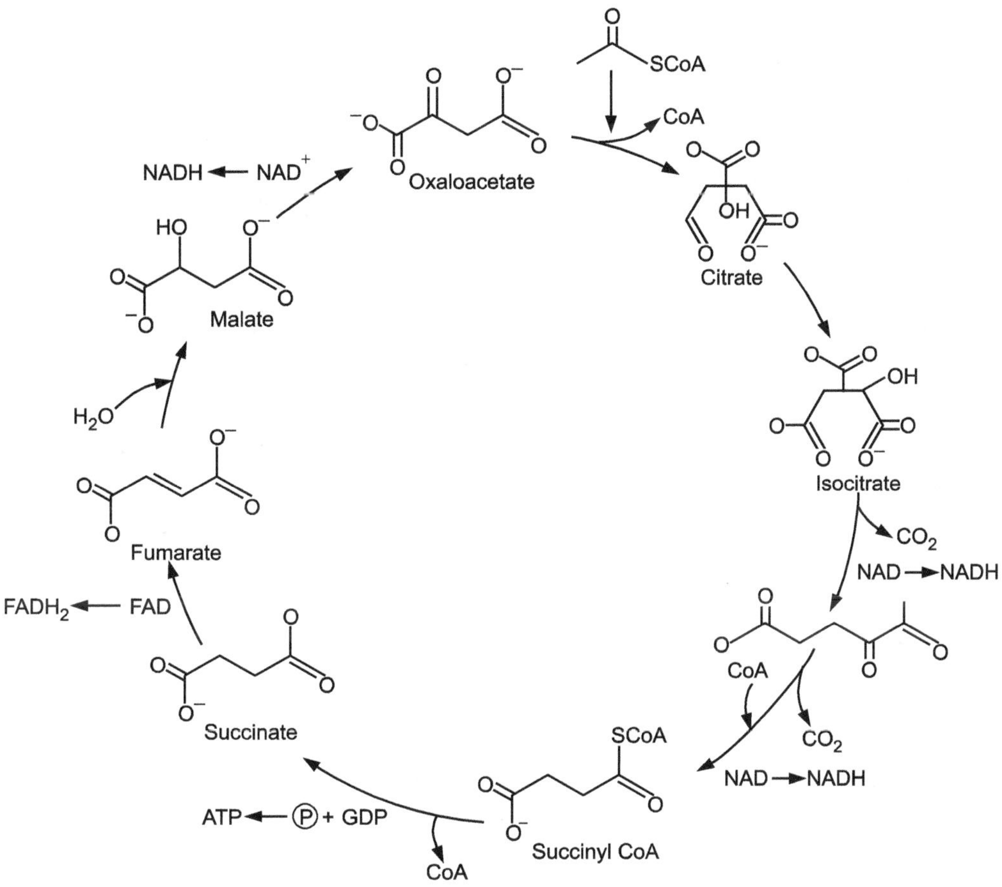

Fig.1.3 TCA Cycle

1.1.4   Pentose Phosphate Pathway

The pentose phosphate pathway (Fig.1.4), also called the phosphogluconate pathway or hexose monophosphate shunt, is a process that generates NADPH and 5-carbon sugars, pentoses. This pathway is an alternative to glycolysis. There are two distinct phases in the pathway. The first is the oxidative phase, in which NADPH is generated, and the second is the non-oxidative synthesis of pentoses. The role of this pathway can be summarized as:

- Production of NADPH,

- Production of ribose-5-phosphate used in the synthesis of nucleotides and nucleic acids.

- Production of erythrose-4-phosphate used in the shikimic acid pathway, synthesis of aromatic amino acids.

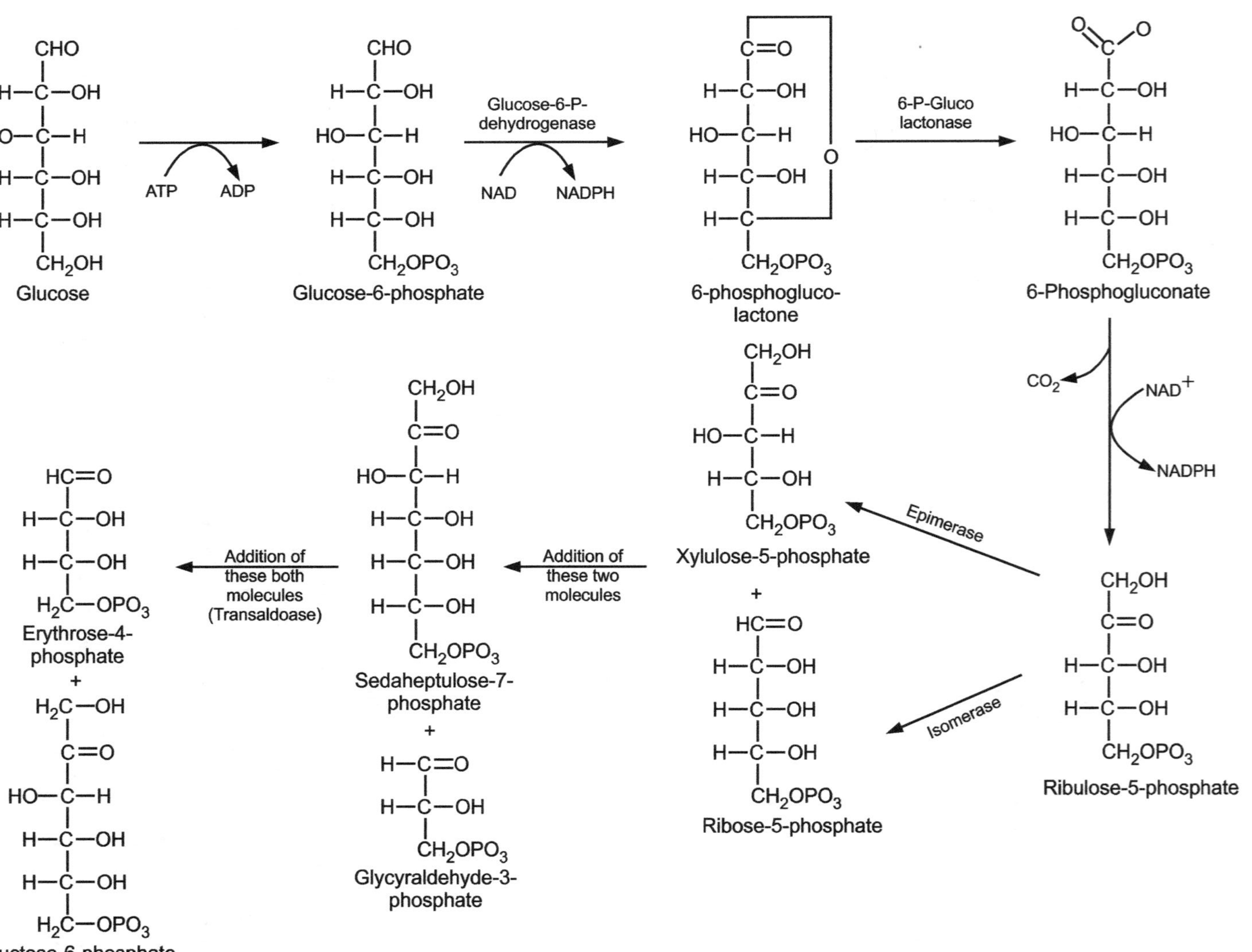

Fig.1.4 Pentose Phosphate Pathway

1.2 Acetate Pathway

1.2.1   Introduction

Acetate pathway is the pathway where acetate unit is the precursor for the biosynthesis of fatty acids and anthracene glycosides. In few compounds mevalonic acid or shikimic acid pathway intermediates are associated with acetate which further produces modified isoprenoidor flavonoid like compounds respectively.

Fatty acids are carboxylic acids with long hydrocarbon chains. The hydrocarbon chain length may vary from 10-30 carbons (most usual is 12-18). Fatty acids, also called as *aliphatic acids,* are important sources of energy stored in the form of triglycerides and act as intermediates in the biosynthesis of polyketides and hormones. Commercially, fatty acid and their derivatives are useful in the manufacturing of food, cosmetics and toiletries products such as soaps, papers, plastic, varnishes, paints and insecticides.

Fatty acids can be saturated and unsaturated depending on double bonds. Saturated fatty acids arealong-chain carboxylic acids that usually contain 12 and 24 carbon atoms with no double bonds. Unsaturated fatty acids which may be mono- or poly-unsaturated, are similar to saturated fatty acids, except that one or more alkenyl functional groups exist along the chain. Monounsaturated fatty acids (MUFAs) have only one double bond. Polyunsaturated fatty acids (PUFAs) have more than one double bond. Fatty acids are frequently represented by a notation such as C18:2 that indicate that the fatty acid consists of an 18-carbon chain and 2 double bonds.

Table 1.1 Fatty Acid ContainingPlants

Plant	Biological Source	Constituents	Uses
Almond	Seeds of *Prunusamygdalus*, Rosaceae	Oleic, linoleic, palmitic, stearic.	Adjuvant, emollient base
Arachis	Seeds of *Arachishypogaea*, Leguminosae	Arachidic, oleic, linoleic, palmitic, stearic	Emollient base
Bees wax	Honey comb of *Apismellifeca*, Apidae	Myristic, palmitic, cerotic	Ointment base
Carnauba	Leaves of *Coperniciaprunifera*, Palmae	Carnaubic, cerotic, melissyl	Opharmaceutical base
Castor	Seeds of *Ricinuscommunis*, Euphorbiaceae	Ricinoleic, oleic, linoleic, palmitic, stearic.	Purgative, emollient base
Chaulmoogra	Seeds of *Hydnocarpusheterophylla*, Flacourtricaceae	Chaulmoogric acid, hydnocarpic acid	Anti T. B., anilprotic
Coconut	Seeds kernel of *Cocosnucifera*, Arecaceae	Lauric, myristic, oleic, palmitic, stearic	Cosmetic preparation
Cod-liver	Fresh liver of *Gadusmorrhua*, Gadidae	Oleic, myristic, palmitic, stearic, DHA	Nutritive supplement
Corn	Grains of *Zeamays*, Gramineae	Oleic, linoleic, palmitic, stearic, arachidic, linolenic	Solvent for injection, cosmetic preparation

Table 1.1 *contd...*

Plant	Biological Source	Constituents	Uses
Cotton seed	Seeds of *Gossypiumhirsutum*, Malvaceae	Oleic, linoleic, palmitic, stearic	Cosmetic preparation
Evening primrose	Seeds of *Oenotherabiennis*, Onagraceae	Oleic, linoleic, palmitic, gamma linolenic.	Nutritive supplement
Jojoba	Seeds of *Simmondsiachinensis*, Buxaceae	Eicosenoic. Docosenoic, oleic acid	Cosmetic preparation
Karanja	Seeds of *Pongamiaglabra*, Papilionaceae	Oleic, linoleic, palmitic, stearic, arachidic, linolenic	Skin diseases
Kokum butter	Seeds of *Garciniaindica*, Guttiferae	Oleic, linoleic, palmitic, stearic, capric	Demulcent, emollient
Lard	Adnominal fat of *Susscrofa*, Suideae	Oleic, linoleic, palmitic, stearic.	Nutritive supplement
Linseed	Seeds of *Linum usitatissimum*, Linaceae	Oleic, linoleic, palmitic, stearic.	Edible oil, demulcent
Mustard	Seeds of *Brassica nigra*, Cruciferae	Arachidic, linolenic, linoleic, oleic, myristic	Counter irritant, rubefacient.
Olive	Fruits of *Olea europaea*, Oleaceae	Oleic, linoleic, palmitic, stearic	Edible oil, emollient base
Palm kernel	Kernel of *Elaeisguineensis*, Arecaceae	Oleic, linoleic, palmitic, stearic, lauric, myristic	Cosmetic toiliteries preparation
Rapeseed	Seeds of *Brassica napus*, Cruciferae	Oleic, linoleic, palmitic, stearic, erucic, alpha linolenic	Edible oil
Rice bran	Seeds of *Oryza sativa*, Gramineae	Oleic, linoleic, palmitic, stearic	Edible oil
Safflower	Seeds of *Carthamus tinctorius*, Compositae	Oleic, linoleic, palmitic, stearic, arachidic, linolenic	Edible oil
Sesame	Seeds of *Sesamum indicum*, Pedaliaceae	Oleic, linoleic, palmitic, stearic	Edible oil, cosmetic toiliteries preparation
Shark liver	Fresh liver of *Hypoprion*brevirostris,	DHA (docosahexaeonic acid) and EPA (eicosapentaenoic acid).	Antixeropthalmic factor
Soya bean	Seeds of *Glycine max*, Leguminosae	Oleic, linoleic, palmitic, stearic, alpha linolenic	Edible oil
Spermaceti	Head of *Physeter macrocephalus*, Physeteridae	Lauric, myristic, stearic, cetyl palmitate	Ointment base
Suet	Abdominal fat from *Oviesaries*, Bovidae	Oleic, myristic, palmitic, stearic	Nutritive supplement
Theobroma	Kernels of *Theobroma cacao*, Sterculiaceae	Oleic, linoleic, palmitic, stearic	Suppository base
Wheat germ	Wheat germs of *Triticum aestivum*, Gramineae	Oleic, linoleic, linolenic	Nutritive supplement
Sunflower	Seeds of *Helianthus annuus*, Compositae	Oleic, linoleic, palmitic, stearic	Edible oil

Table 1.2 Saturated Fatty Acids Examples

Common Name	Structural Formula
Propionic acid	CH_3CH_2COOH
Butyric acid	$CH_3(CH_2)_2COOH$
Valeric acid	$CH_3(CH_2)_3COOH$
Enanthic acid	$CH_3(CH_2)_5COOH$
Caprylic acid	$CH_3(CH_2)_6COOH$
Capric acid	$CH_3(CH_2)_8COOH$
Lauric acid	$CH_3(CH_2)_{10}COOH$
Myristic acid	$CH_3(CH_2)_{12}COOH$
Palmitic acid	$CH_3(CH_2)_{14}COOH$
Margaric acid	$CH_3(CH_2)_{15}COOH$
Stearic acid	$CH_3(CH_2)_{16}COOH$
Arachidic acid	$CH_3(CH_2)_{18}COOH$
Behenic acid	$CH_3(CH_2)_{20}COOH$
Tricosylic acid	$CH_3(CH_2)_{21}COOH$

Table 1.3 Unsaturated Fatty Acids Examples

Common Name	Chemical Structure
Myristoleic acid	$CH_3(CH_2)_3CH=CH(CH_2)_7COOH$
Palmitoleic acid	$CH_3(CH_2)_5CH=CH(CH_2)_7COOH$
Oleic acid	$CH_3(CH_2)_7CH=CH(CH_2)_7COOH$
Linoleic acid	$CH_3(CH_2)_4CH=CHCH_2CH=CH(CH_2)_7COOH$
α-Linolenic acid	$CH_3CH_2CH=CHCH_2CH=CHCH_2CH=CH(CH_2)_7COOH$
Arachidonic acid	$CH_3(CH_2)_4CH=CHCH_2CH=CHCH_2CH=CHCH_2CH=CH(CH_2)_3COOH$
Eicosapentaenoic acid	$CH_3CH_2CH=CHCH_2CH=CHCH_2CH=CHCH_2CH=CHCH_2CH=CH(CH_2)_3COOH$
Erucic acid	$CH_3(CH_2)_7CH=CH(CH_2)_{11}COOH$
Docosahexaenoic acid	$CH_3CH_2CH=CHCH_2CH=CHCH_2CH=CHCH_2CH=CHCH_2CH=CHCH_2CH=CH(CH_2)_2COOH$

1.2.2   Saturated Fatty Acid Biosynthesis

Saturated fatty acids are synthesized by a series of decarboxylative Claisen condensation reactions from acetyl-CoA and malonyl-CoA in the presence of enzyme fatty acid synthase. Enzyme synthase contains ACP as part of its structure. Following each step of elongation the β-keto group is reduced to the fully saturated carbon chain by the sequential action of enzymes – ketoreductase, dehydratase and enol reductase. Fatty acid synthesis (Fig.1.5) starts with acetyl CoA which is the two carbon containing precursor. This is used to add two carbons to growing fatty acid chain stepwise. This explains why fatty acids always have an even number of carbons. This process occurs in the cytosol.

During biosynthesis, the growing fatty acid chain gets attached covalently to the phosphopantethiene prosthetic group of ACP (acyl carrier protein) which allows intermediates to remain covalently linked to the synthases and access this intermediates to the right enzyme-active sites. Acyl Carrier Protein (ACP), converts malonyl CoA to malonyl ACP. ACP synthase is an enzyme that holds the growing fatty acid chain. Acyl enzyme thioester releases C_2 unit to malonyl ACP to form fatty acyl ACP through sequence of reduction, dehydration reactions. This fatty acyl ACP on attack of water generates fatty acids. Triglycerides (esters of glycerol containing same or different 3 fatty acids) are biosynthesized from glycerol 3-P (product of Calvin cycle) and first fatty acyl CoA esterification process which is elaborated in Fig.1.5.

Fatty acid synthesis is simply a linear combination of acetate units facilitated by enzyme fatty acid synthase. ACP allows growing fatty acid chain to react with thio group of enzyme fatty acid synthase and thus head to tail condensation followed by reduction gives rise to a long chain of saturated fatty acid. Mostly even numbers of carbon containing fatty acids are common in nature. But when starting compound is other than acetate (example- propionic acid), odd number of C-containing fatty acids can also be synthesized by plants. C_{16} and C_{18} (Palmitic and stearic acids respectively) are the most common saturated fatty acids.

In fatty acid biosynthesis, acetate is the starter group and malonate is a chain extender. But in a few compounds there may be change in starter or chain extender group. Cinnamoyl CoA obtained from shikimic acid pathway acts as a starter group in the synthesis of flavonoid and stilbenes. Anthranilolyl CoA obtained from anthranilic acid is used in the synthesis of quinoline and acridine alkaloids. Hexanoate is the starter group in the formation of aflatoxins and cannabinoides. Incorporation of propionate as a chain extender other than mevalonate from propionyl CoA or methyl malonyl CoA leads to the formation of macrolide antibiotics.

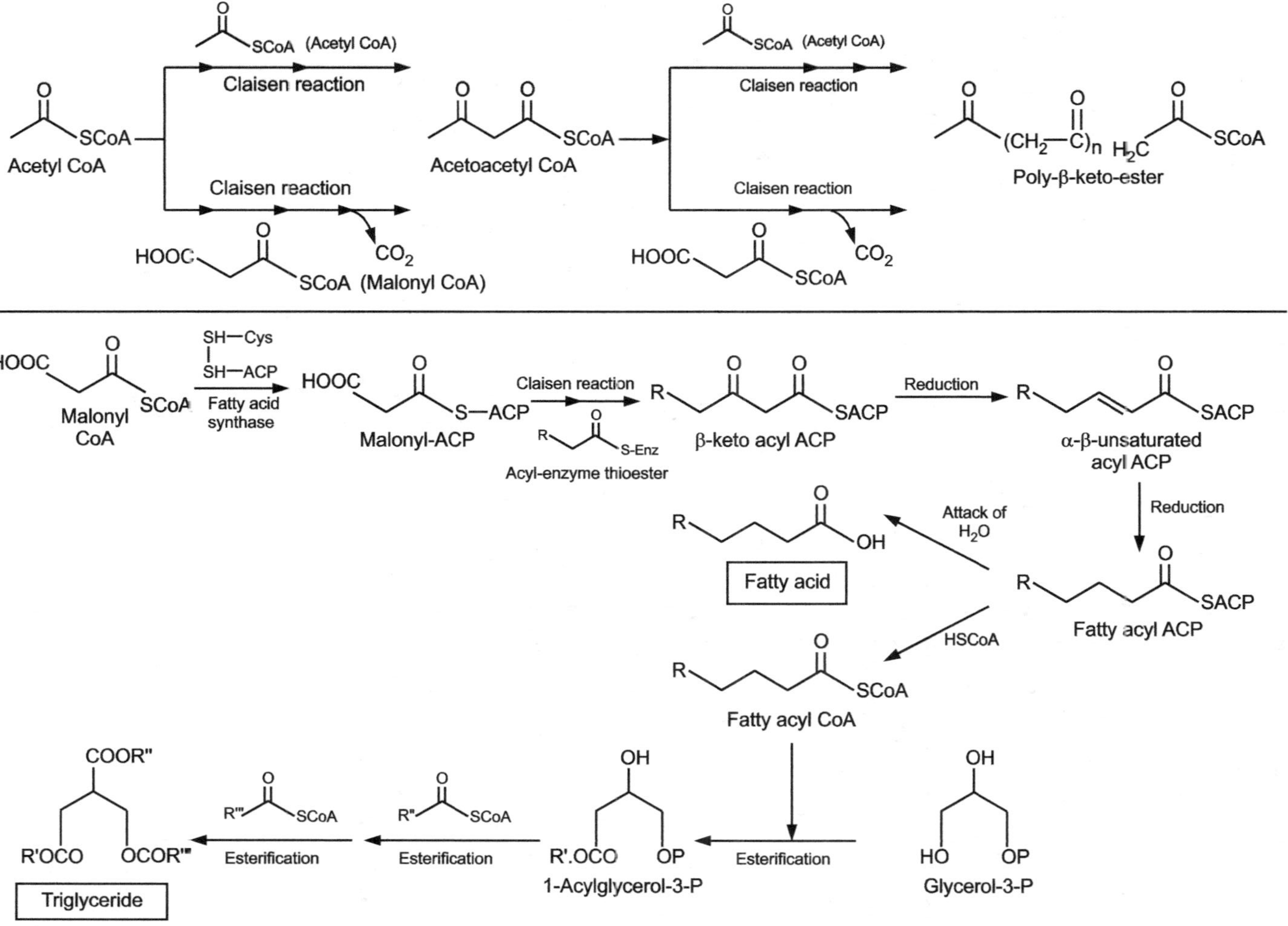

Fig.1.5 Acetate Pathway

1.2.3   Unsaturated Fatty Acid (UFA) Biosynthesis

UFAs are synthesized by sequential desaturation mechanism. Double bond is introduced by enzyme desaturase, oxidation and NADPH or NADH. Stearic acid is the common starting material to give various UFAs.

Essential fatty acids belong to the class of PUFAs. There are two types of essential fatty acids; omega-6 and omega-3 fatty acids which can be short chain (omega-3α linolenic acid, omega-6 eicosapentaenoic acid or EPA, docosahexaenoic acid DHA) or long-chain polyunsaturated fatty acids (omega-3α Linoleic acid, omega-6α gamma-linolenic acid or GLA, dihomo-gamma-linolenic acid or DGLA,   arachidonic acid or AA). Linoleic acid and isomers of linolenic acid are EFA. As its derivative linolenate blocks synthesis of prostaglandins and hence useful in the treatment of various ailments especially cardiovascular diseases. Linoleic acid found to be the important compound in the prevention of hair loss, cancer, cystic fibrosis and dermatitis. DHA is present in mothers's milk and constituent of brain and retina. The consumption of alpha linolenic acid can be converted to DHA by animals which is used to prevent brain related diseases, colon cancer and cardiovascular risks.

In animals, EFA are obtained from diet whichact as precursors for synthesis of DHA, which is (Fig.1.6) important (component) for normal functioning of brain. Dihomolinoleic acid (PGE$_1$), arachdinoic acid (PGE$_2$) and EPA (PGE$_3$) are important precursors for biosynthesis of prostaglandins. PGH$_2$ synthesizes thromboxanes and arachidonic acid syntehsizes leukotrienes. Eicosanoid is the collective term for oxygenated derivatives of three different 20-carbon essential fatty acids i.e.eicosapentaenoic acid (EPA), arachidonic acid (AA) and dihomo-gamma-linolenic acid (DGLA). There are four types of eicosanoids i.e. prostaglandins, prostacyclins, thromboxanes and leukotrienes.

Various branched chain fatty acids are biosynthesized by various mechanisms. Sterculi, malvalic acid and chaulmoogric acid are some of the important branched chain fatty acids obtained from stearic acid by alkylation, desaturation (sterculic) oxidation (malvalic). 2-cyclopentenyl carboxyl CoA as a starter unit and malonyl CoA as an extender unit gives chaulmoogric acid.

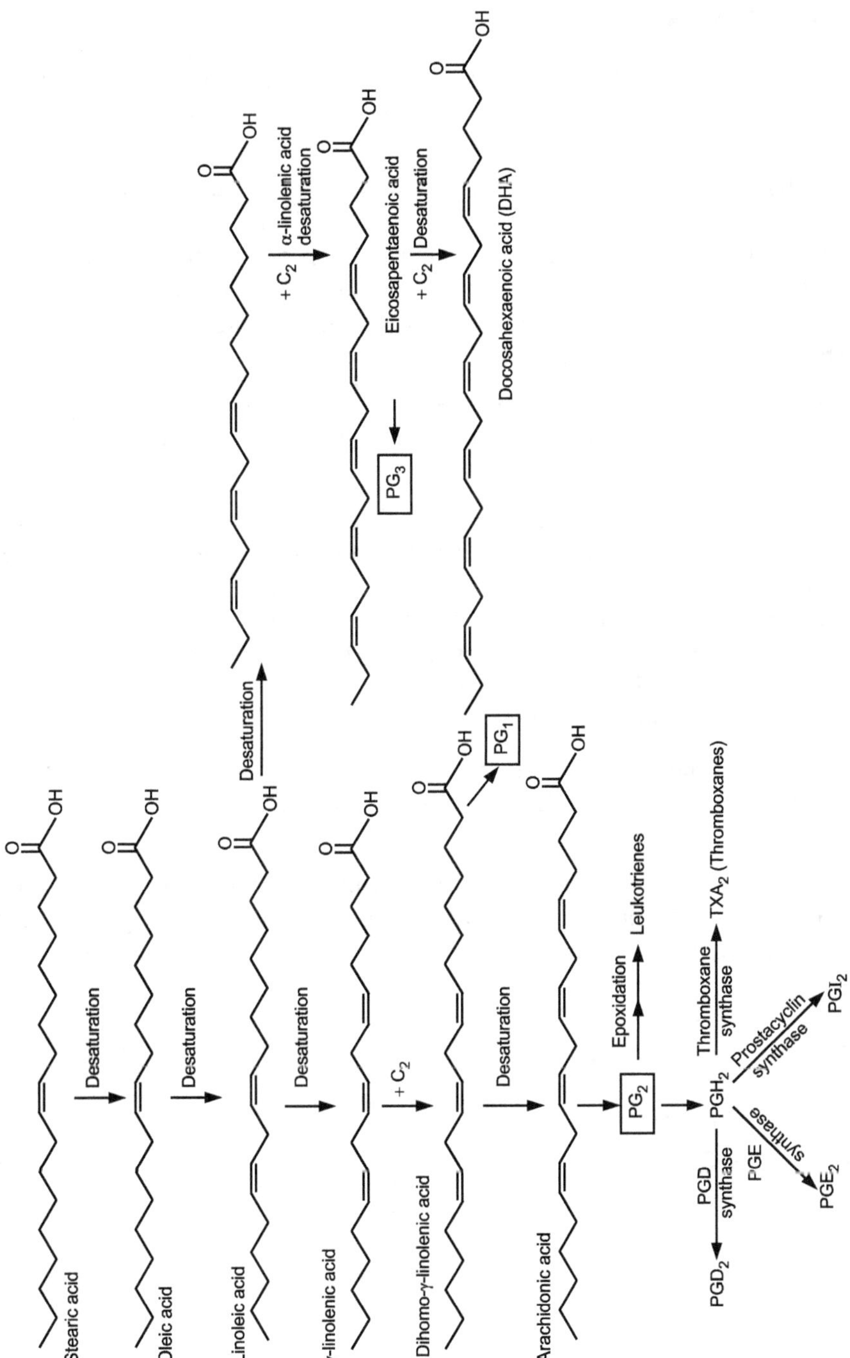

Fig.1.6 Biosynthesis of Prostaglandins, thromboxanes and Leukotrienes - an overview and PUFA

1.3   Shikimic Acid Pathways

Shikimic acid pathway is significantly important as it leads to formation of almost all aromatic compounds present in nature like phenylpropanoides (lignans, lignnins), coumarin, flavonoids and isoflavonoids and terpenoid quinones.

Shikimate pathway (Fig.1.7) starts with erythrose 4-phospate (obtained from the pentose phosphate pathway) and phosphoenolpyruvate (obtained from glycolysis pathway) coupling to yield phosphorylated 7-carbon keto sugar, 3-deoxy-D-arabino-heptulosonic acid-7-phosphate, (DHAP). DHAP on removal of phosphoric acid cyclizes to 3-dehydroquinic acid which on reduction yields quinic acid. By dehydration3-dehydroquinic acid forms 3 dehydroshikimic acid which forms shikimic acid followed by reduction.

3-dehydroshikimic acid on dehydration and enolisation forms protocatechuic acid and on dehydrogenation and enolisation forms gallic acid which is the component of many types of tannins.

Shikimic acid through phosphorylation and elimination reactions forms very important intermediate chorismic acid. Chorismic acid via simple rearrangement gives prephenic acid. Chorismic acid on another branch by glutamine mediated amination at C_2 gives anthranilic acid while amination at C_4 position (Fig. 1.8) gives P-aminobenzoic acid (PABA). PABA is a part of folic acid structure. Prephenic acid on dehydration and decarboxylation yields precursor of phenylalanine i.e.phenylpyruvic acid. On dehydrogenation and decarboxylation, prephenic acid yields p-hydroxy phenylpyruvic acid which is direct precursor of tyrosine.

Isochorismic acid, isomer of chorismic acid, on pyruvic acid elimination forms salicylic acid known as phenolic phytohormone present in *Salix alba*, Salicaceae and *Filipendulaulmaria*, Rosaceae. (Fig.1.8).

Phenylalanine on enzymatic deamination forms cinnamic acid which is the starting material for biosynthesis of various phenylpropanoids. Phenylpropanoid compounds are so named because of the basic structure of a three-carbon side chain on an aromatic ring, which is derived from L-phenylalanine.

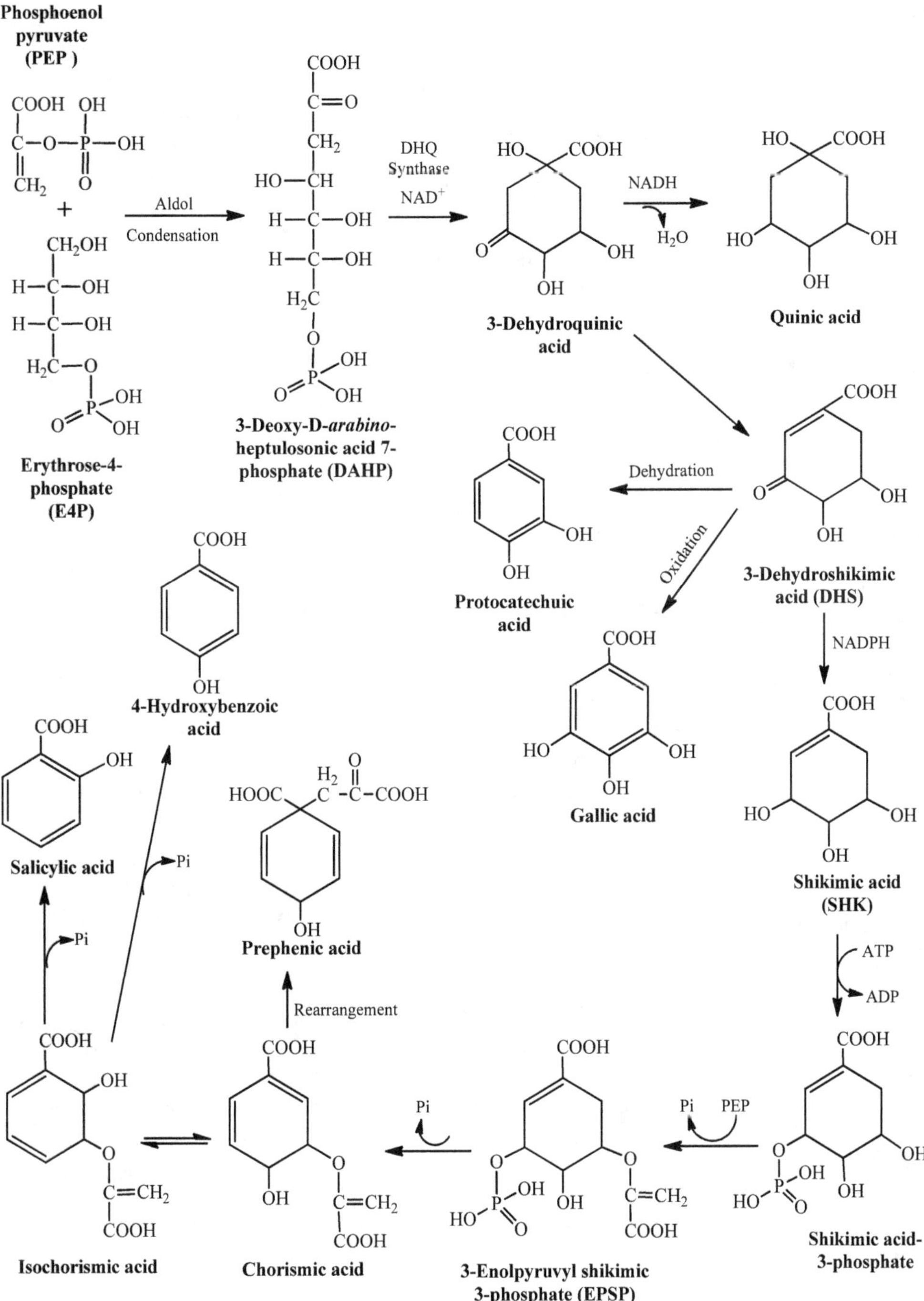

Fig.1.7 Shikimic acid Pathway Part-I

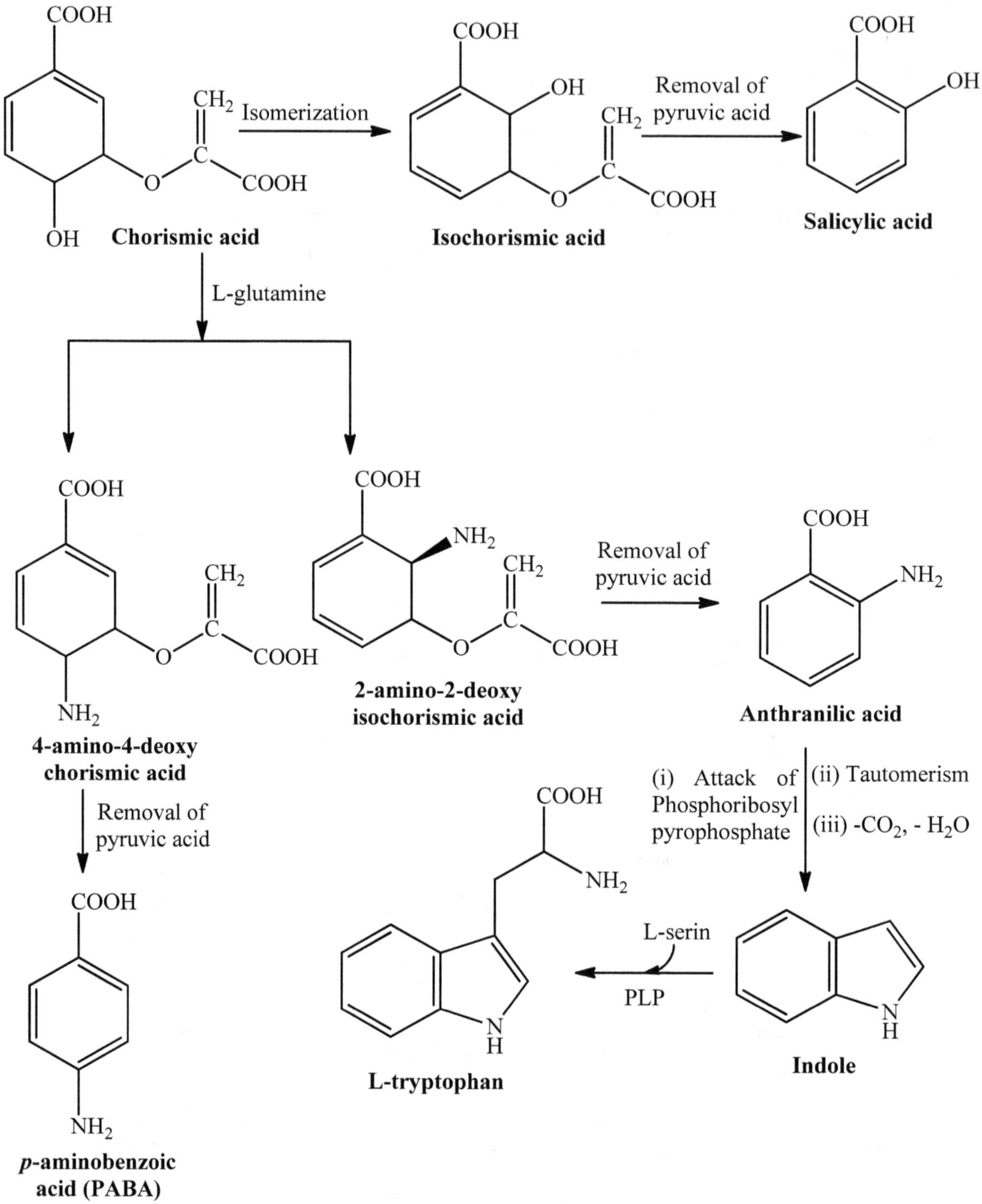

Isomerization

Removal of pyruvic acid

Chorismic acid

Isochorismic acid

Salicylic acid

L-glutamine

Removal of pyruvic acid

4-amino-4-deoxy chorismic acid

2-amino-2-deoxy isochorismic acid

Anthranilic acid

Removal of pyruvic acid

(i) Attack of Phosphoribosyl pyrophosphate

(ii) Tautomerism

(iii) $-CO_2$, $- H_2O$

L-serin

PLP

Indole

L-tryptophan

***p*-aminobenzoic acid (PABA)**

Fig.1.8 Shikimic acid Pathway–Part-II

1.4   Amino Acid Biosynthesis Pathway

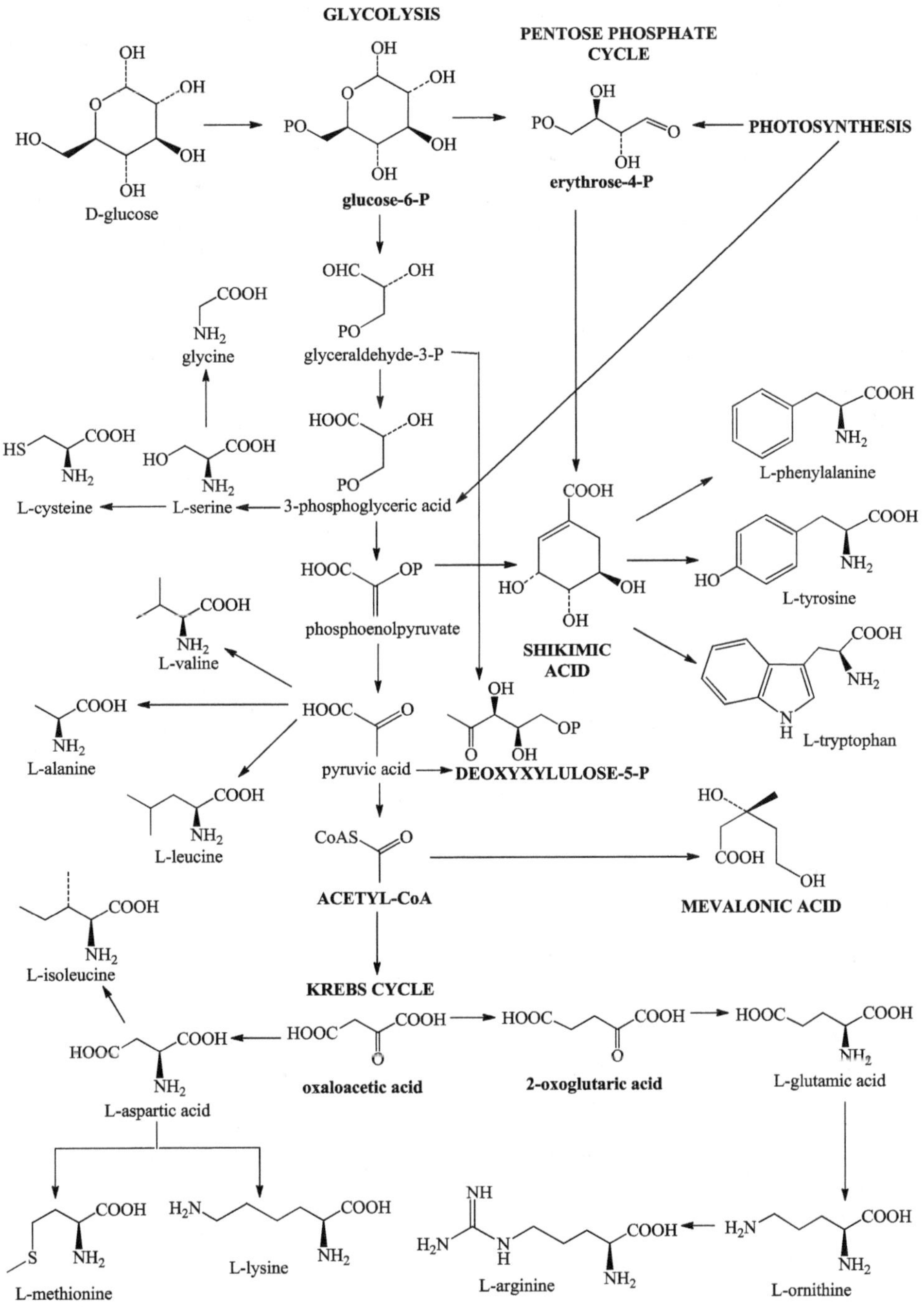

Fig.1.9 General scheme of Amino acid biosynthesis

- All amino acids are derived from intermediates in glycolysis, the citric acid cycle, or the pentose phosphate pathway (Fig.1.9).

- Nitrogen enters these pathways by way of glutamate and glutamine. Some pathways are simple, others are not.Biosynthesis of serine, glycine, cysteine, homocysteine, methionine, proline, orinithine, arginine are outlined in Fig. 1.10 to 1.14.

- Ten of the amino acids are only one or a few enzymatic steps removed from their precursors. The pathways for others, such as the aromatic amino acids, are more complex.

- Different organisms vary greatly in their ability to synthesize the 20 amino acids. Whereas most bacteria and plants can synthesize all 20, mammals can synthesize only about half of them

4-hydroxybenzoic acid

Chorismic acid

L-Gln *amination using ammonia (generated from glutamine) as nucleophile* L-Gln

isochorismic acid

salicylic acid

hydrolysis of enol ether side-chain

4-amino-4-deoxy-chorismic acid

2-amino-2-deoxy-isochorismic acid

NAD^+

2,3-dihydroxybenzoic acid

p-aminobenzoic acid (PABA)

anthranilic acid

L-tryptophan

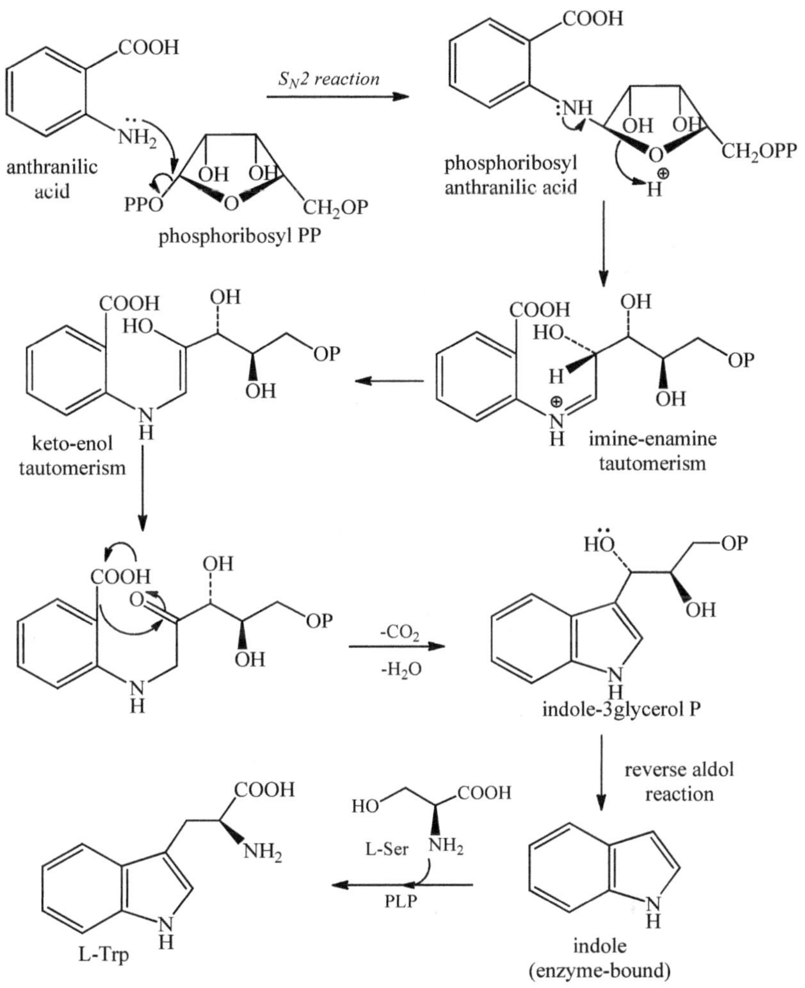

Fig.1.10 Biosynthesis of Tryptophan

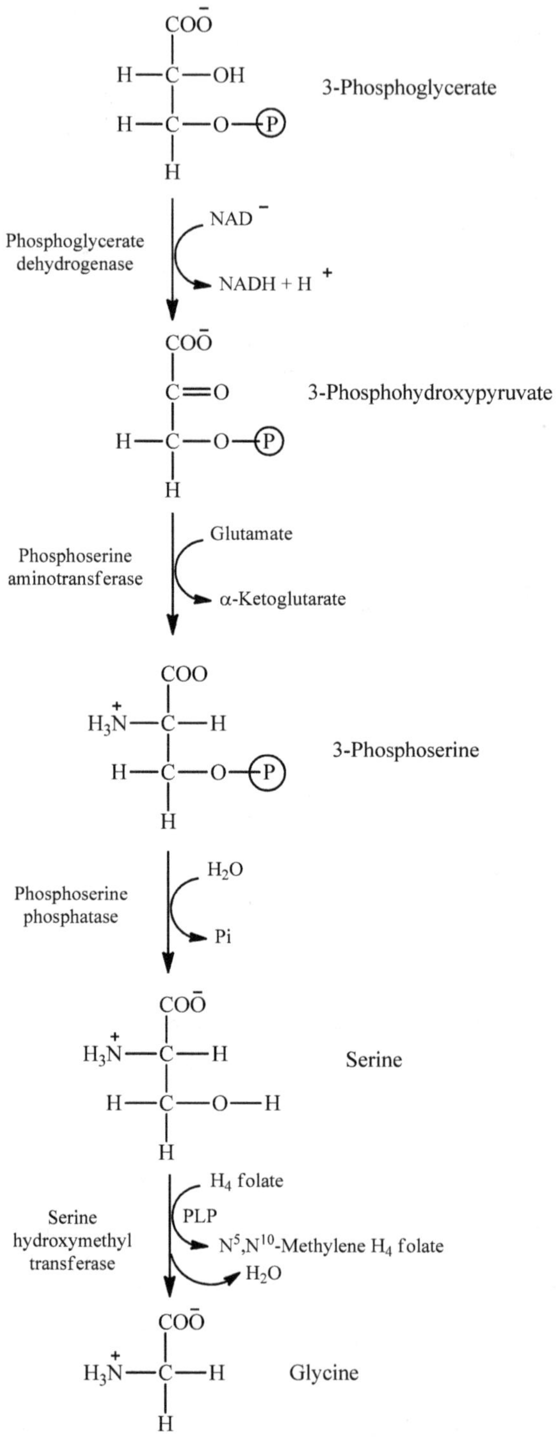

3-Phosphoglycerate

Phosphoglycerate dehydrogenase — NAD^- → $NADH + H^+$

3-Phosphohydroxypyruvate

Phosphoserine aminotransferase — Glutamate → α-Ketoglutarate

3-Phosphoserine

Phosphoserine phosphatase — H_2O → Pi

Serine

Serine hydroxymethyl transferase — H_4 folate, PLP → N^5,N^{10}-Methylene H_4 folate, H_2O

Glycine

Fig.1.11 Biosynthesis of Serine, Glycine

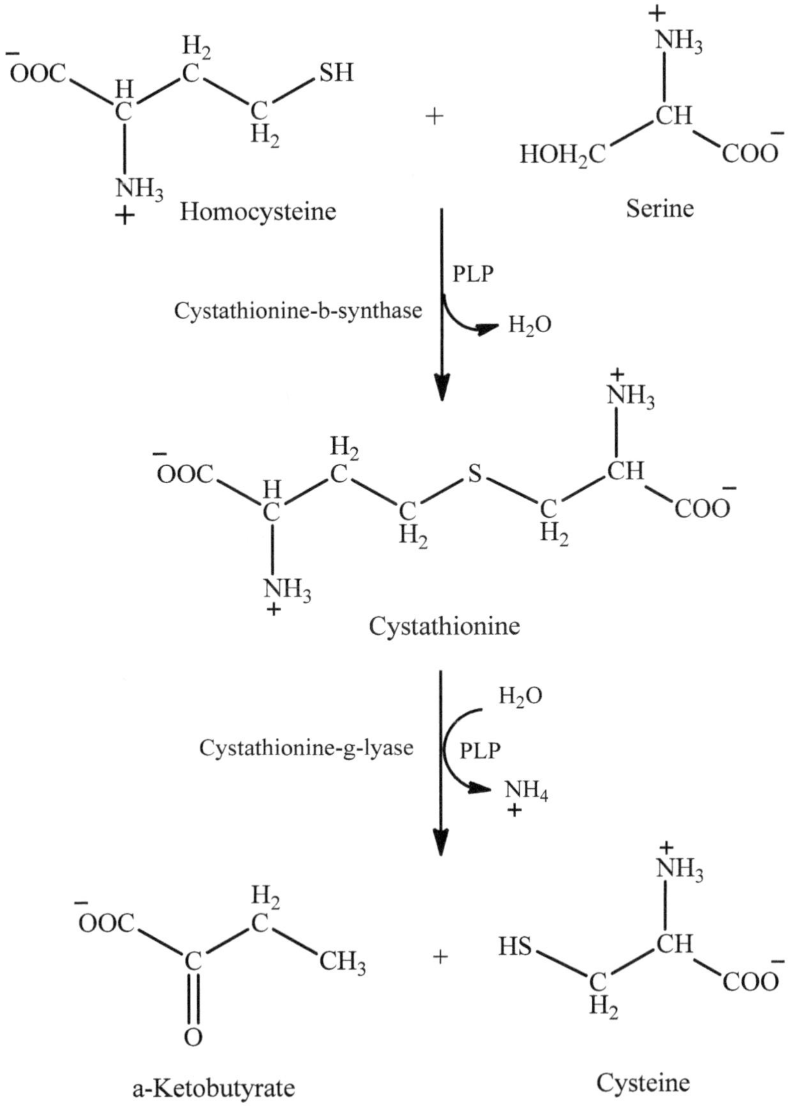

Fig.1.12 Biosynthesis of Cysteine

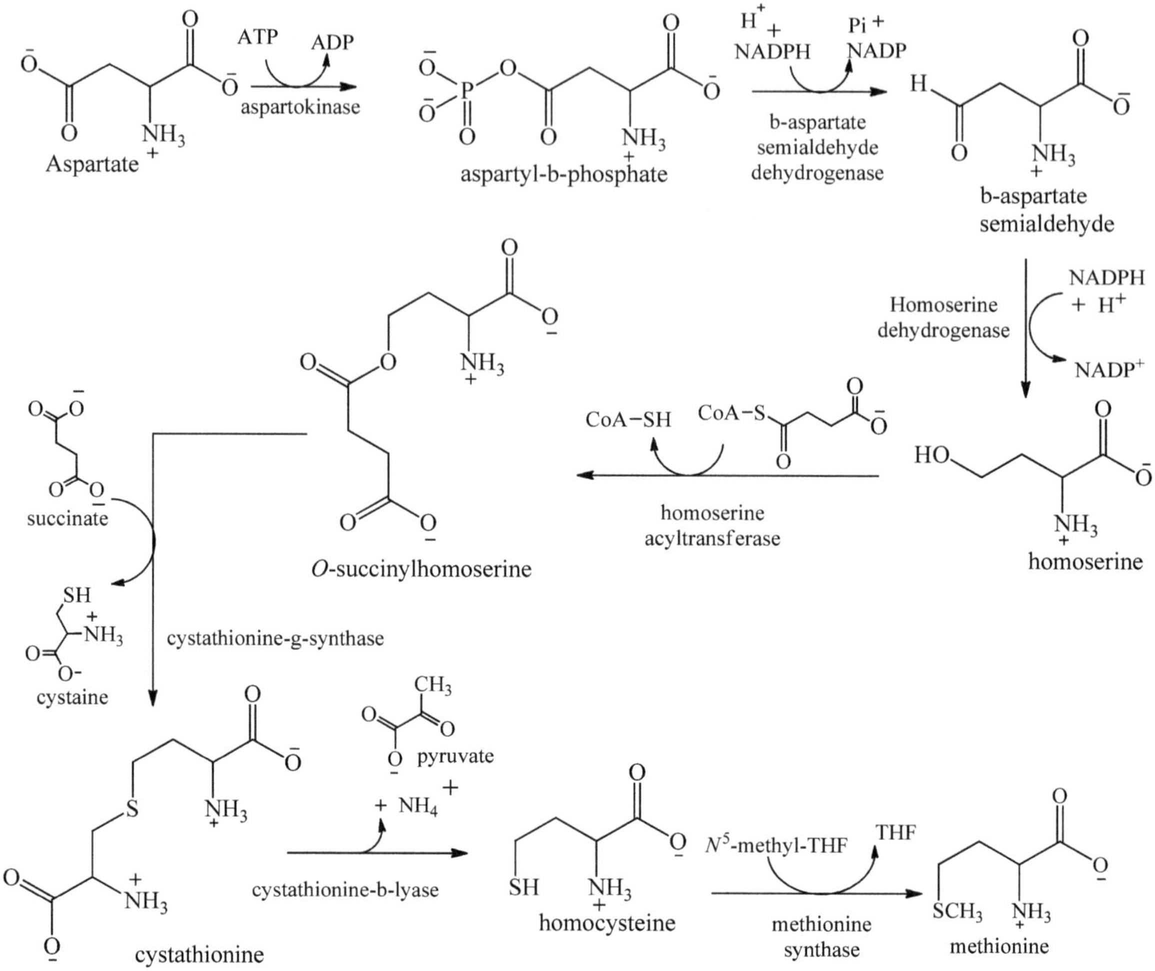

Fig.1.13 Biosynthesis of homocysteine and methionine

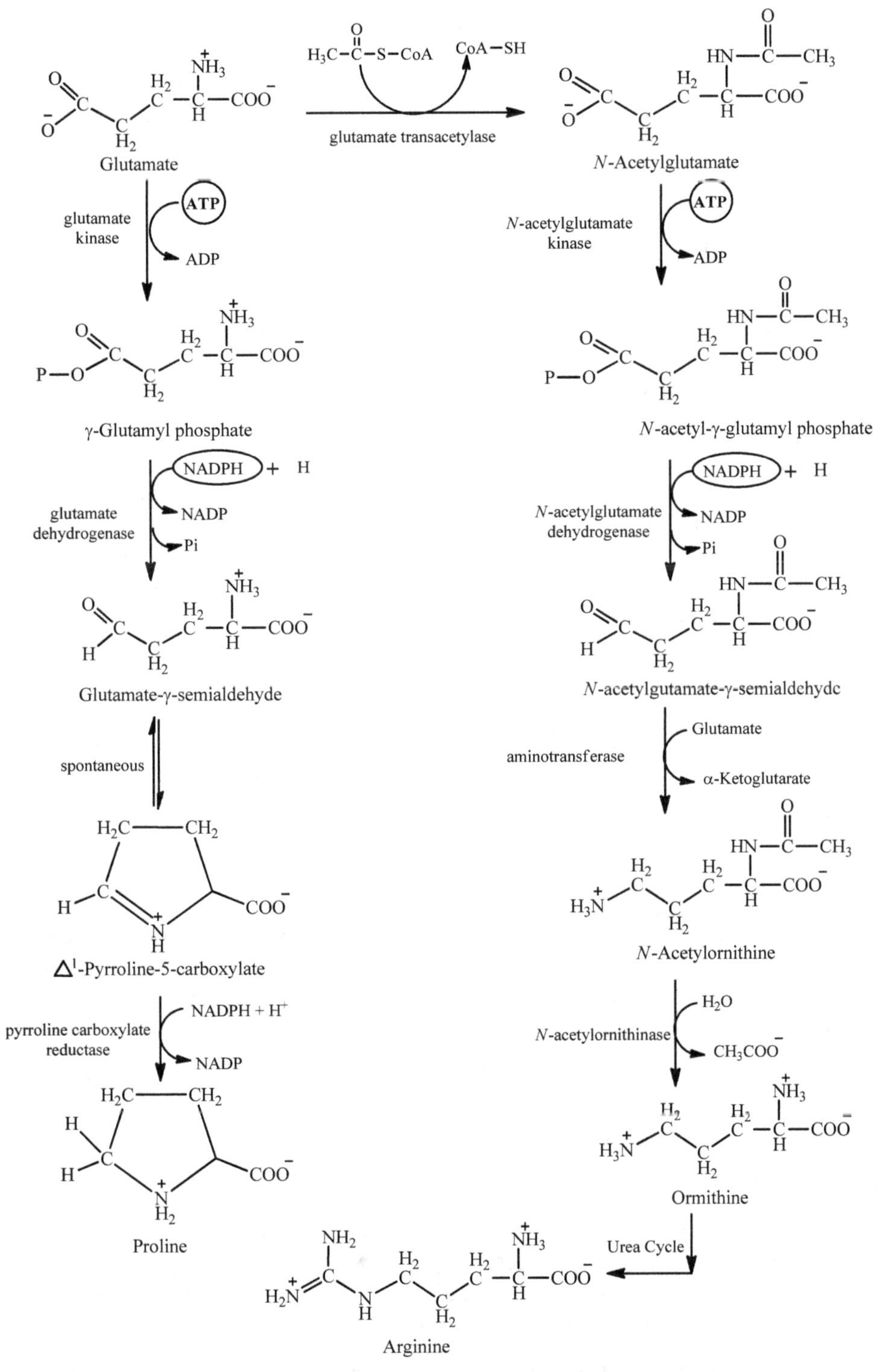

Glutamate

glutamate transacetylase

N-Acetylglutamate

glutamate kinase — ATP → ADP

N-acetylglutamate kinase — ATP → ADP

γ-Glutamyl phosphate

N-acetyl-γ-glutamyl phosphate

glutamate dehydrogenase — NADPH + H → NADP → Pi

N-acetylglutamate dehydrogenase — NADPH + H → NADP → Pi

Glutamate-γ-semialdehyde

N-acetylgutamate-γ-semialdehyde

spontaneous

aminotransferase — Glutamate → α-Ketoglutarate

Δ^1-Pyrroline-5-carboxylate

N-Acetylornithine

pyrroline carboxylate reductase — NADPH + H$^+$ → NADP

N-acetylornithinase — H$_2$O → CH$_3$COO$^-$

Proline

Ormithine

Urea Cycle

Arginine

Fig.1.14 Biosynthesis of Proline, Orinithine, Arginine

1.5 Elucidation of Biosynthetic Pathway

After discussing the above things now one question will certainly rise that how this pathways of secondary metabolite biosynthesis are elucidated? Actually the elucidation of pathway is very first step of biosynthesis study. New advances in chemistry and analytical techniques have made possible to explore various biosynthetic pathways with a very clear picture of precursors, intermediates, products, enzymes and reactions involved.

Techniques of Elucidation of Biosynthetic Pathway

Technique	Details
Use of isolated organ	It is useful to locate site of biosynthesis of particular compounds. This is Tissue culture technique useful to determine site of synthesis as well as whole pathway through addition of radioactive tracers in cultures of plants parts. Example: *Datura* and *Nicotinana* species
Grafting method	It is also useful to locate site of biosynthesis of particular compounds. In this method, cut portion of one plant (scion) is grafted on major plant (stock) to identify and confirm exact site of synthesis of secondary metabolites. Example: scion of datura plant and stock of tomato plant showed less accumulation of tropane alkaloids while stock of datura plant and scion of tomato plant increased amount of alkaloids.
Mutant strain use	It is useful to confirm intermediates and enzymes responsible for biosynthesis of particular compounds. Natural or induced mutation in lower plants or fungi useful to determine and confirm intermediates or product due to blockages of pathway. Many times one of the intermediates get accumulated and thus whole biosynthesis stopped that time artificial supply of enzymes or intermediate helps in normalizing pathway.
Enzyme studies	It is useful to confirm enzymes responsible for reactions. Various enzymes can be isolated and studied for their exact role in different biosynthetic reactions by traditional biochemical methods.
Genomic studies	It is useful to study chemical reactions involved in biosynthesis of metabolites via responsible proteins and or related DNA and genes by simple biochemical studies or modern genetic analysis
Tracer technique	It involves use of radioactive and stable isotopes to elucidate biosynthetic pathways. Details are as follows:

Study of utilization of radioactive isotopes in the investigation of Biogenetic studies

(Tracer Technique)

Use of radio labelled tracers is the most preferred technique of elucidation due to its specificity and selectivity. Tracer technique is also useful to confirm the exact precursor or intermediate from the number of possible precursors or intermediates. The whole sequence can also be generated by use of tracer technique. It involves the following steps:

Step-1 Selection of suitable radioisotope: The selection of radioisotope depends on half-life so that it should withstand the long and unpredictable time period of biosynthesis. ^{14}C and ^{3}H are the most commonly used radio isotopes. Now-a-days, stable isotopes like ^{13}C are also utilized to elucidate biosynthesis with the help of sophisticated NMR analysis. Stable isotope labelling involves the use of non-radioactive isotopes that can act as a tracers used to model several chemical and biochemical systems. The chosen isotope can act as a label on that compound that can be identified through nuclear magnetic resonance (NMR) and mass spectrometry (MS).

Properties of radioactive and stable isotopes		
Radioactive isotope	**Radiation**	**Half-life**
Carbon (C^{14})	Beta	5760 years
Hydrogen (H^3)	Beta	12.5 years
Sulphur (S^{35})	Beta	871 days
Phosphorus (P32)	Beta	14.3 days
Chlorine (Cl^{36})	Beta	4.4×10^5 years
Iodine (I^{131})	Beta, gamma	8 days
Cobalt (Co^{60})	Beta, gamma	5.3 days

Step-2 Labelling of precursor or intermediate: The labelling of any metabolite by radioisotope is a tedious and complex procedure involving many chemical steps

Step-3 Insertion of radio-labelled metabolite in plant part: After labelling the desired metabolite, the next step is to introduce it into the suitable organ of a plant (e.g. root, stem, leaves) by proper method (e.g. immersion, injection and spraying). Tissue cultures of plant parts are also preferred for introduction of labelled metabolite/s. The whole pathway elucidation as well as the exact site of biosynthesis can be studied by tissue culture. Intact roots can be grown in tracer solution. Even the plant can be grown in an atmosphere containing $^{14}CO_2$ or $^{13}CO_2$ tracers. An example of palladium-catalyzed halogen exchange is shown below for the synthesis of tritium-labelled mephenytoin. Here the parent molecule is iodinated with N-iodosuccinimide (NIS) in triflic acid, followed by Pd-catalyzed iodine exchange to give the labelled compound Here the parent molecule is iodinated with N-iodosuccinimide (NIS) in triflic acid, followed by Pd-catalyzed iodine exchange to give the labelled compound [3H3](S)-mephenytoin.

Example of radio labelling

Step-4 Time-to-time determination of radio-activity: After sufficient period of time the metabolite or intermediate or products are isolated to check the radioactivity by suitable detectors. Example: Geiger-Muller Counter, Scintillation Counter. The modern unit of radioactivity is Becquerel which means onedisintegration per second (dps). The specific detection of radioactive isotopes is straightforward; with liquid scintillation counting (LSC) being the main method by which long-lived radiolabels is quantitatively detected. LSC uses a photomultiplier tube to detect light emissions from the fluor; a fluor is a fluorescent molecule that undergoes excitation by the absorption of radiation and releases light when it relaxes to the ground state. The amount of light emitted by a specified amount of radioactive material can be directly correlated to the amount of radioactivity present.

Step-5 Establishment of precursor-inter-mediate-productrelation: If radioactivity is detected then the precursor-product relationship can be confirmed. Similarly, each step should be elucidated and enzymes catalyzing these steps should be identified.

Various tracer Technique Methods

Tracer technique methods	General Procedure	Example
Precursor Product Sequence	To feed presumed precursor into the plant material, isolate and study radioactivity after pre-decided time to establish precursor product sequence	Restricted synthesis of hyoscine, distinct from hyoscyamine in *Datura stramonium*
	 Fig.1.15 Precursor Product Sequence	
Double and Multiple Labelling	To feed double or multiple labelled precursor to know exact which nitrogen or hydrogen is involved in biosynthesis of product	Doubly labelled lysine used to determine which hydrogen of lysine molecule was involved in formation of piperidine ring of anabasine in *Nicotina glauca*.
Competitive Feeding	To feed two or more precursors to confirm exact precursor-intermediate-product sequence	Biosynthesis of different tropane alkaloids
	 Fig.1.16 Example of competitive feeding	

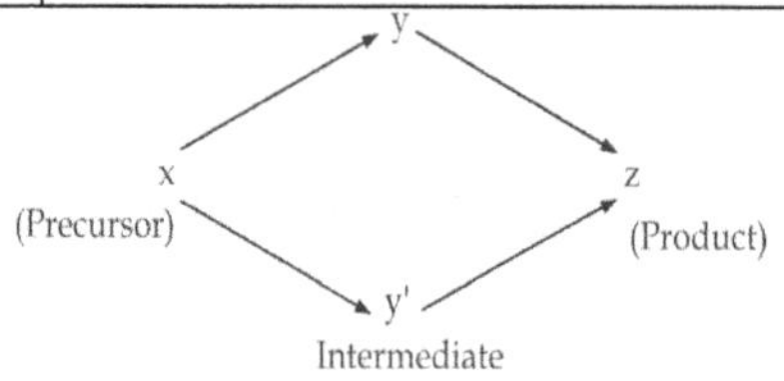

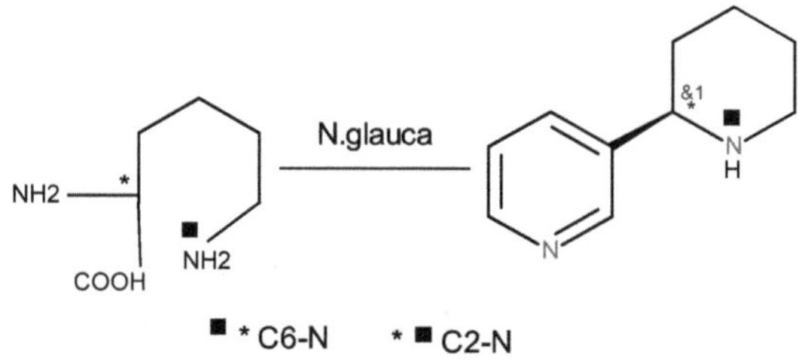

For the Precursor Product Sequence figure:

For the competitive feeding figure:

Tracer technique methods	General Procedure	Example
Isotope Incorporation	To feed different isotopes to confirm position of bond cleavage and formation	Glucose – 1- phosphatase cleavage as catalyzed by alkaline phosphatase
Sequential Analysis	To grow plant in atmosphere of $^{14}CO_2$ and then analyze the plant at given time interval to obtain the sequence in which various correlated compound become labelled.	Determination of sequential formation of opium alkaloids

Subjective Questions

1. Discuss shikimic acid biosynthesis pathways with its significance.
2. Discuss Acetate pathway with its significance.
3. Discuss aromatic amino acid pathway with its significance.
4. How glutamic acid and alanine are biosynthesized?
5. How proline, arginine and orinithine are biosynthesized.
6. How to elucidate biosynthesis pathways by tracer technique?
7. What are different techniques of elucidation of biosynthetic pathways?
8. Which alkaloids are biosynthesized from aromatic amino acids?
9. Why radioactive isotopes are useful in elucidation of biosynthetic pathways?

Multiple Choice Questions (MCQs)

1. Which one of the following technique is used for elucidation of biosynthetic pathway in plants?
   a. X – Ray diffraction
   b. Electron microscope
   c. Tracer technology
   d. None

2. Which labelled isotope is used specifically to understand biosynthesis of alkaloids, proteins and amino acids?
   a. Nitrogen
   b. Sulphur
   c. Carbon
   d. Phosphorus

3. Shikimic acid is synthesized from ....
   a. Hexose phosphate
   b. Pentose phosphate
   c. Heptose phosphate
   d. None

4. Shikimic acid is key intermediate for biosynthesis of ....

   a. $C_6 - C_3$ units

   b. $C_6 - C_4$ units

   c. $C_3 - C_6$ units

   d. $C_5 - C_3$ units

5. Which one of the following is precursor of shikimic acid?

   a. Chorismic acid

   b. Pyruvic acid

   c. Phosphoenol pyruvic acid

   d. None

6. Which one of the following is immediate precursor of tyrosine in shikimic acid pathway?

   a. P-phenyl pyruvic acid

   b. Prephenic acid

   c. P- hydroxy phenyl pyruvic acid

   d. Chorismic acid

7. Isoprenoids and steroids are biosynthesized by.................

   a. Acetate mevalonate pathway

   b. Both a and b

   c. Shikimic acid pathway

   d. None

8. Fatty acids are synthesized from.............

   a. Acetate Mevalonate pathway

   b. Both a and b

   c. Acetate Mevalonate pathway

   d. None

9. C_5 compound isopentenyl pyrophosphate is derived from mevalonic acid pyrophosphate by

   a. Dehydration and decarboxylation

   b. Dehydration

   c. Decarboxylation and dehydration

   d. Decarboxylation

10. Monoterpenes (C_{10}units) are synthesized from ....

    a. Farnesyl pyrophosphate

    b. Geranyl pyrophosphate

    c. Isopentenyl pyrophosphate

    d. None

11. C_{15} unit isoprenoid compounds are synthesized from..............

    a. Farnesyl pyrophosphate

    b. Geranyl pyrophosphate

    c. Isopentenyl pyrophosphate

    d. None

12. Squalene is precursor of ....

    a. Steroids

    b. Diterpenes

    c. Triterpenes

    d. Both a and b

13. Choose the correct sequence involved in cholesterol synthesis from the following

    a. Acetate ,Malonte, isopentenyl pyrophosphate, Squalene pathway

    b. Acetate, Mevalonate , isopentenyl pyrophosphate, Squalene pathway

    c. Acetate , Mevalonate, pentenyl pyrophosphate , Squalene pathway

    d. Acetate, Mevalonate , isopentenyl phosphate , Squalene pathway

14. Choose the correct sequence

a. Shikimic acid	Chorismic acid	Anthranilic acid	Tryptophan
b. Shikimic acid	Prephenic acid	Anthranilic acid	Tryptophan
c. Shikimic acid	Chorismic acid	Anthranilic acid	Tryptophan
d. Shikimic acid	Chorismic acid	Phenyl pyruvic acid	Tryptophan

15. Choose the correct sequence

a. Shikimic acid	Prephenic acid	Phenyl pyruvic acid	Phenylalanine
b. Shikimic acid	Chorismic acid	Phenyl pyruvic acid	Phenylalanine
c. Shikimic acid	Chorismic acid	Phenyl pyruvic acid	Tyrosine
d. Shikimic acid	Prephenic acid	Phenyl pyruvic acid	Tyrosine

16. The radioisotopes used for tracing the biosynthetic pathways are.............

    a. $^{14}C, ^{3}H, ^{36}S, ^{32}P$ b. $^{15}C, ^{1}H, ^{34}S, ^{34}P$

    c. $^{13}C, ^{2}H, ^{34}S, ^{30}P$ d. $^{11}C, ^{2}H, ^{33}S, ^{31}P$

17. Which .... amino acid is derived from shikimic acid pathway?

    a. Tyrosine   b. Isoleucine

    c. Valine   d. Phenylanine

18. Malonyl CoA is precursor of ....

    a. Lipids   b. flavonoids

    c. Steroids   d. Glycosides

19. IUPAC name of mevalonic acid is........

    a. 2,3,5 – trihydroxy – 3- methylavaleric acid

    b. 5-hydroxy-3-methylavaleric acid

    c. 3-hydroxy -3-methylavaleric acid

    d. 3, 5 – dihydroxy -3-methylavaleric acid

20. Which one of the following is structure of shikimic acid?

a.   c.

b.

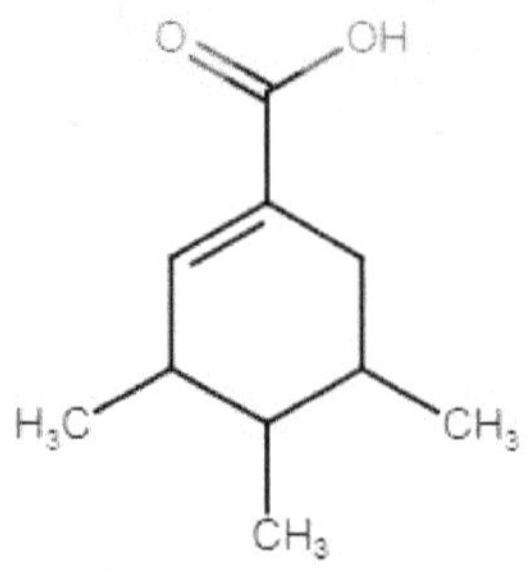

d.

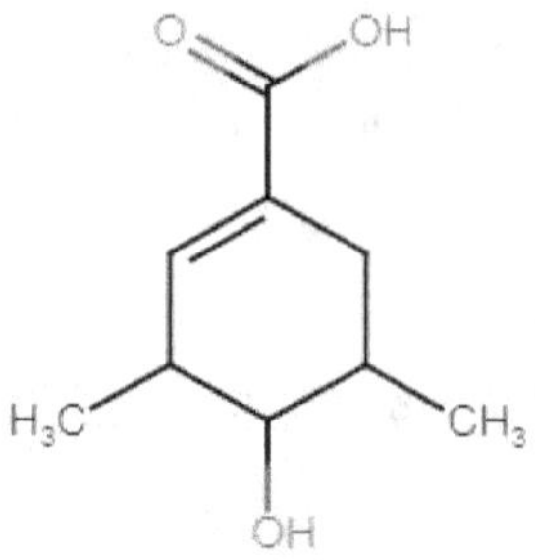

Answer Key

1. c	2. a	3. c	4. a	5. b	6. b	7. a	8. a	9. b	10. C
11. a	12. d	13. a	14. b	15. a	16. a	17. b	18. a	19. d	20. a

Unit 2

General Introduction

PCI Syllabus

Generalintroduction, composition, chemistry & chemical classes, biosources, therapeutic uses andcommercialapplicationsof followingsecondary metabolites.

Chapter Content

2.1   **Alkaloids**

2.2   **Phenylpropanoids and Flavonoids**

2.3   **Steroids, Cardiac Glycosides and Triterpenoids**

       2.3.1 Steroids (Steroidial Saponin Glycosides)

2.4   **Volatile Oils**

2.5   **Tannins**

2.6   **Resins**

2.7   **Glycosides**

2.8   **Iridoids, other Terpenoids and Naphthaquinones**

2.1 Alkaloids

Alkaloids are defined as "physiologically active basic compounds of plant origin, in which at least one nitrogen atom forms part of a cyclic system". Majority of alkaloids are of diverse chemical nature with potent therapeutic effects. Nitrogen atom is mostly derived from amino acids, but few of alkaloids are not amino acid derivatives like purine and steroid alkaloids. The main criterion for the classification of alkaloids is the type of fundamental (normally heterocyclic) ring structure present in alkaloids. They are broadly categorised into two divisions: *heterocyclic* (divided into groups according to the nature of their heterocyclic ring) and *non-heterocyclic* in which nitrogen is part of side chain (also called as proto or amine alkaloids).

Properties

Most of the alkaloids are generally colorless (except yellow berberine, red betainidin, copper red sanguinarine, crystalline solids (except nicotine, coniine, sparteine which are liquid in nature). Most of the alkaloids contain one or more nitrogen atoms usually in the tertiary state in a ring system. Alkaloids are optically active and majority being a levorotatory except (+) tubocurarine is more potent than (-) tubocurarine.Thereis considerable difference in pharmacological activities of different isomers of alkaloids. For example:

- ➤ (-) ephedrine more active than (+) ephedrine,
- ➤ (-) ergotamine more active than (+)ergotamine
- ➤ (-) and (+) quinine both are pharmacologically active,

In plants, alkaloids generally exists as salts of organic acid (like acetic acid, oxalic acid, citric acid, malic acid, lactic acid, tartaric acid, tannic acid) due to their basic nature.

- ➤ Opium alkaloids are found in the salt form of meconic acid.
- ➤ Cinchona alkaloids are found with quinic acid.
- ➤ Aconite alkaloids with aconitic acid.

Some alkaloids like narceine and nicotine are occurring free in nature. A few alkaloids also occur as glycosides of sugars like glucose, rhamnose and galactose. For example, *Solanum* species contains glycoalkaloids.

General classification of alkaloids

- ➤ *True alkaloids*: The true alkaloids are basic in nature, contain heterocyclic nitrogen which is derived from amino acids. Example: morphine, emetine, hyoscyamine.

- ➤ *Proto alkaloids*: Proto alkaloidsare called as amino alkaloids or simple amines are also basic in nature, derived from amino acids but nitrogen is in side chain and not the part of cyclic ring. Example: mescaline, colchicines, ephedrine.

Pseudo alkaloids: These alkaloids are not derived from amino acids but shows coloration with the standard qualitative tests for alkaloids. It includes terpenoidal, steroidal and purine alkaloids. Example: caffeine, conessine.

Biosyhtetic classification of alkaloids

Chemical Type	Biosynthetic Origin
Tropane and Pyrrolizidine alkaloids	Ornithine
Piperidine, Quinolizidine, Indolizidine alkaloids	Lysine
Pyridine alkaloids	Niconitic acid
Tetrahydro isoquinoline, Benzyltetra-hydro isoquinoline, Phenthyl isoquinoline alkaloids	Tyrosine
Indole, Quinoline alkaloids	Tryptophan
Quinazoline and Acridine alkaloids	Anthranilic acid
Imidazole alkaloids	Histidine
Amine alkaloids	Phenylalanine
Terpenoid alkaloids	Monoterpenes
Steroid alkaloids	Steroids
Purine alkaloids	Purine

Chemical classification of alkaloids

True Alkaloids	
Fundametal Chemical Moiety	**Examples**
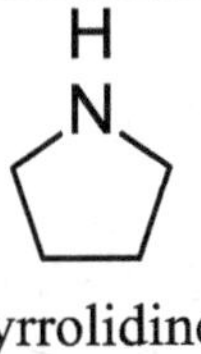 Pyrrolidine	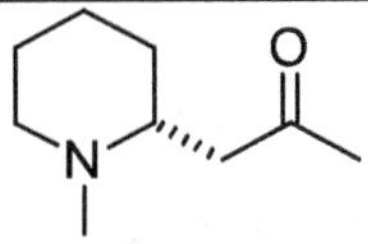 Hygrine, *Erythroxylon coca*, Erythroxylaceae
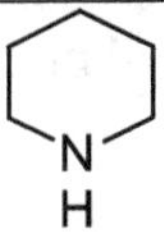 Piperdine	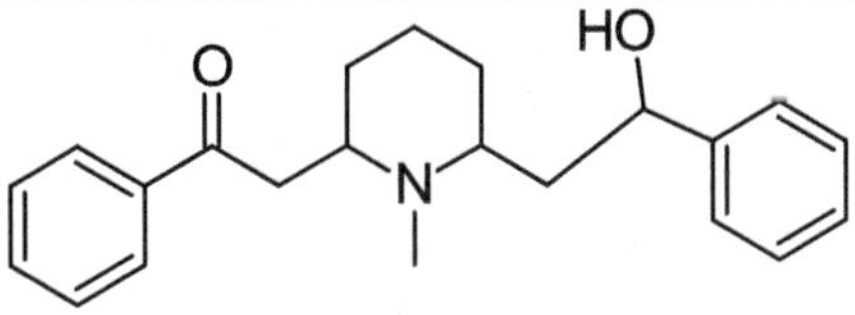 Lobeline, *Lobelia inflata,* Lobeliaceae (Respiratory Stimulant)

Contd...

	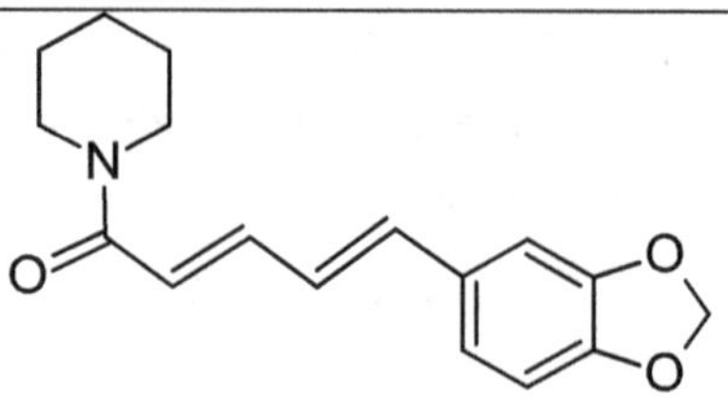 Piperine (Bioavailability enhancer)
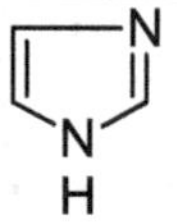 Imidazole	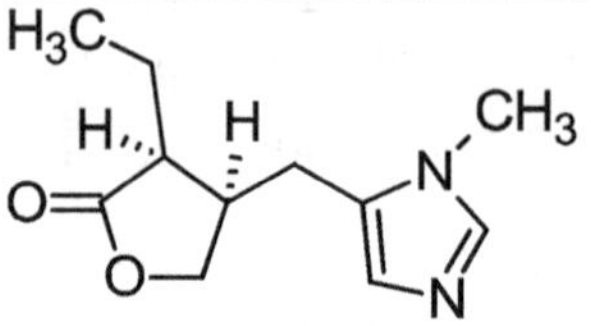 Pilocarpine, *Pilocarpus jaborandi*, Rutaceae To treat dry mouth and eye issues. (Eye drops it is used to manage angle closure glaucoma until surgery can be performed, ocular hypertension, primary open angle glaucoma, and to bring about constriction of the pupil following its dilation.)
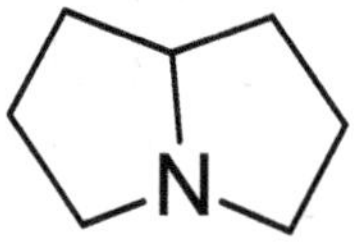 Pyrrolizidine	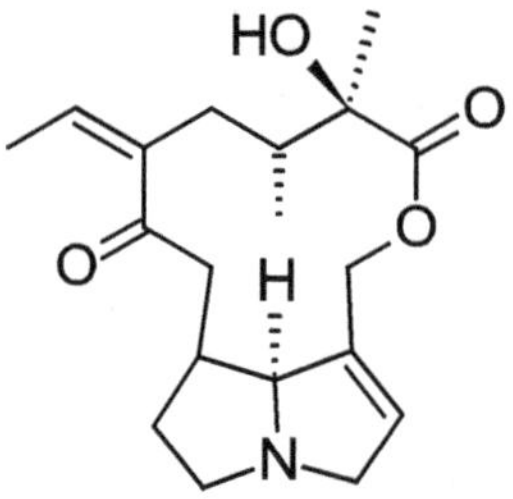 Senecionine, *Senecio brasiliensis*, Asteraceae (It is used as arrow Poision)
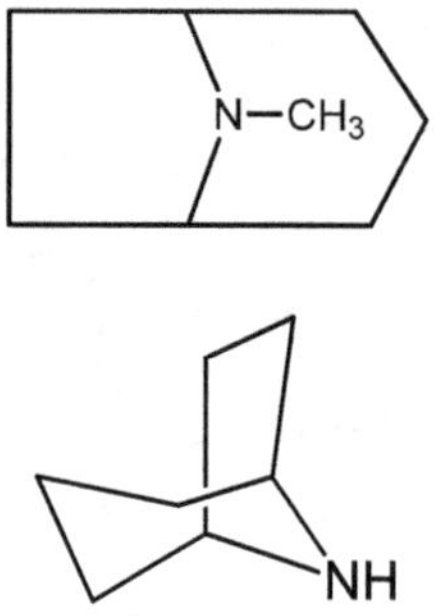 Tropane	Atropine, *Atropa belladonna*, Solanaceae (Atropine is indicated for the treatment of bradycardia associated with hypotension, second- and third-degree heart block, and slow idioventricular rhythms. Atropine is the initial drug of choice in acute organophosphate poisoning.)

Contd...

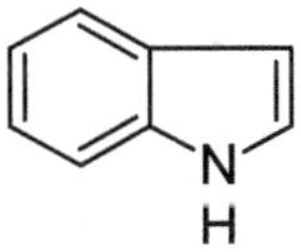 Indole (Benzpyrrole) Vinca, Raulwoflia, Nux vomica	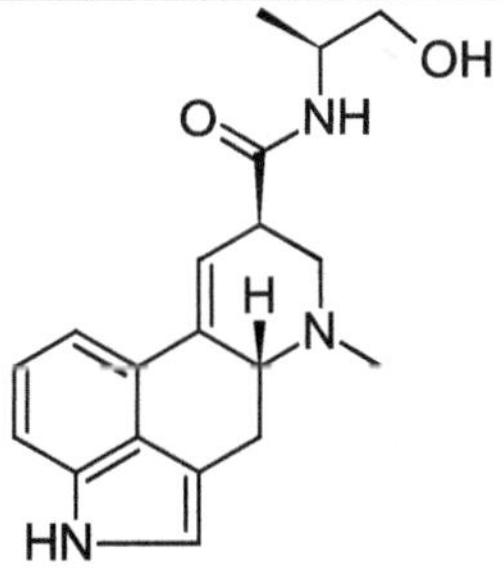 *Ergot alkaloids, Clavicepus purpurea,* Clavicepataceae (Ergometrine is Oxytocic, Ergotamine isuse to treat migraine or cluster headache)
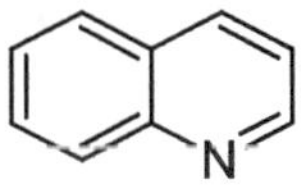 Quinoline	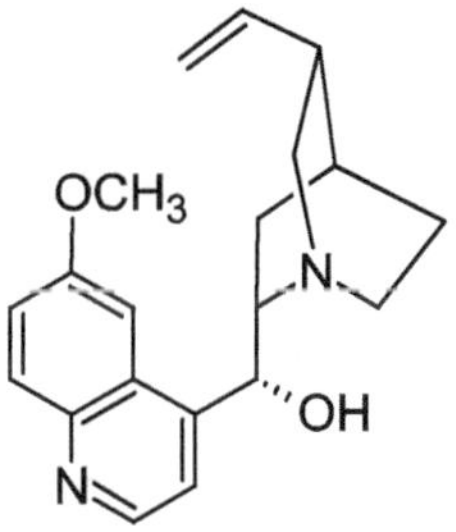 Quinine, *Cinchona officinalis*, Rubiaceae (Anti-malarial)
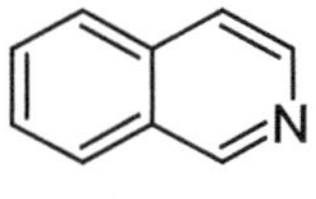 Isoquinoline	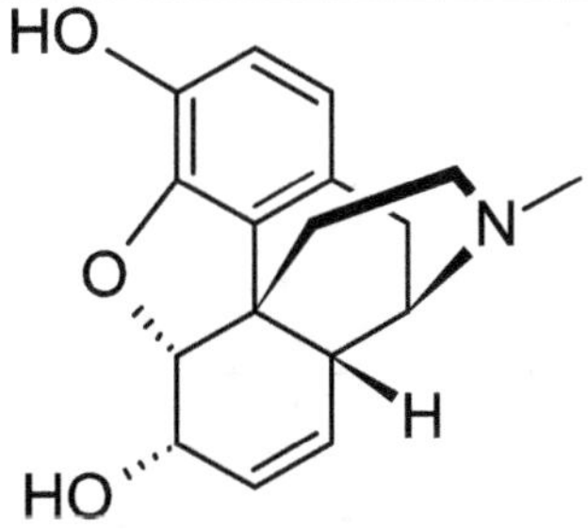 Morphine, *Papaverumsomniferum*, Papaveraceae (Narcotic sedative)
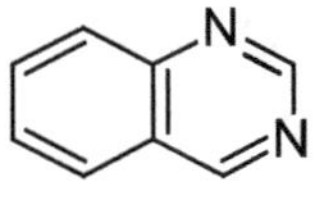 Quinazoline	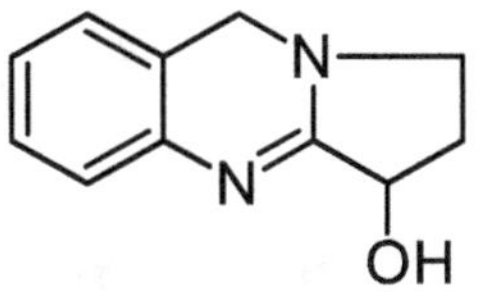 Vasicine, *Adhatodavasica, Acanthaceae* (Expectorant and bronchodilator. weak cardiac stimulant)

Contd...

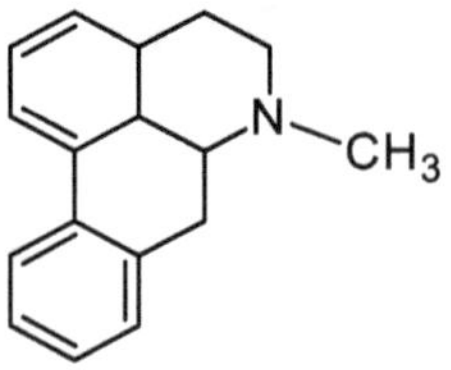 Aporphine	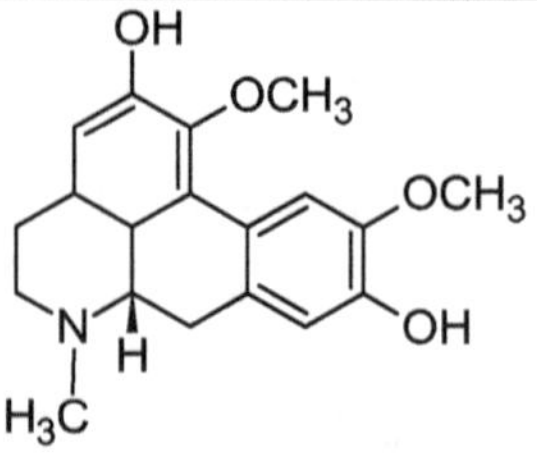 Boldine, *Peumusboldus*, Monimiaceae (Antioxidant,Hepatorpotective, neuroprotective)

Pseudo Alkaloids

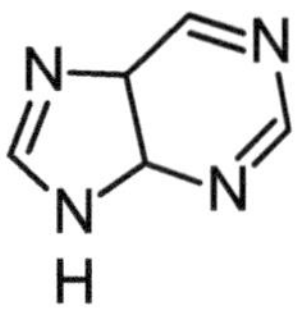 Purine/Xanthine	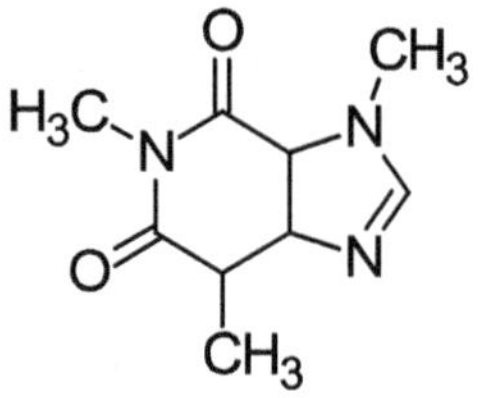 Caffeine, *Thea sinensis*, Theaceae CNS Stimulant
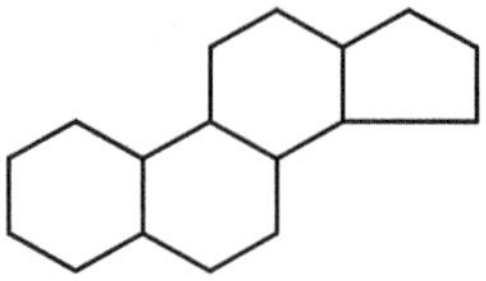 Steroid	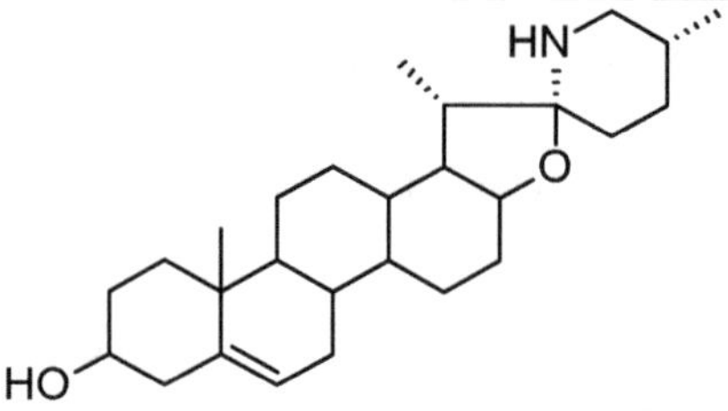 Solasodine, *Solanum species, Solanaceae*
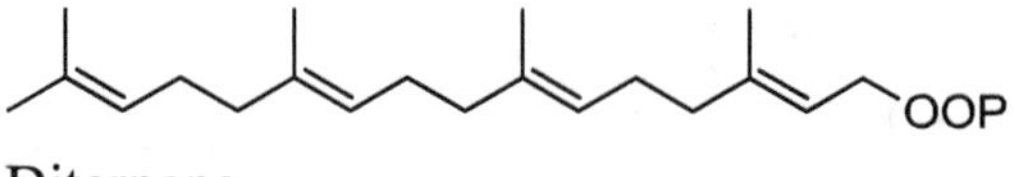 Diterpene	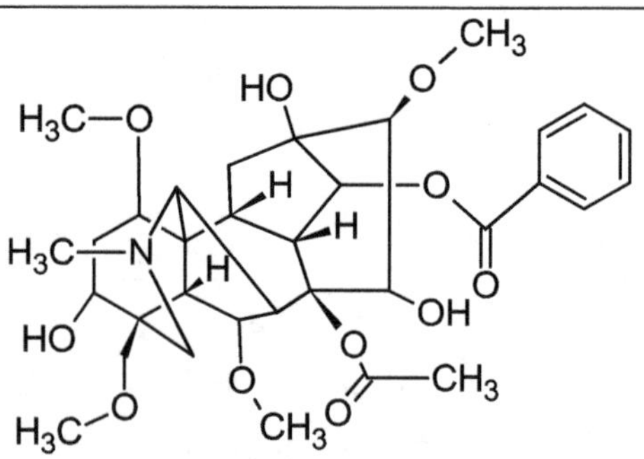 Aconitine *Aconitum napellus*, Ranunculaceae (Potent Poision- 0.028 mg/kg. In small doses as pain relief caused by

Contd...

	trigeminal and intercostal neuralgia, rheumatism, migraine, and general debilitation)
Proto or Amine Alkaloids	
H | R—N—H Amines	Ephedrine from *Ephedra gerardiana*, ephedraceae (Bronchodialator to treat cough)

Extraction of Alkaloid

General Method: Alkaloid extraction process involves frequent treatment with acid and alkali. This is due to the fact that alkaloids are always associated with different acids such as salts and treatment with alkali liberates respective free bases. These free bases are easy to extract in non-polar to semi-polar solvents which avoids extraction of undesirable side products. Addition of acid to this basic solution causes formation of water soluble alkaloid salts. Thegeneral method used for total alkaloid extraction is shown in Fig.2.1.

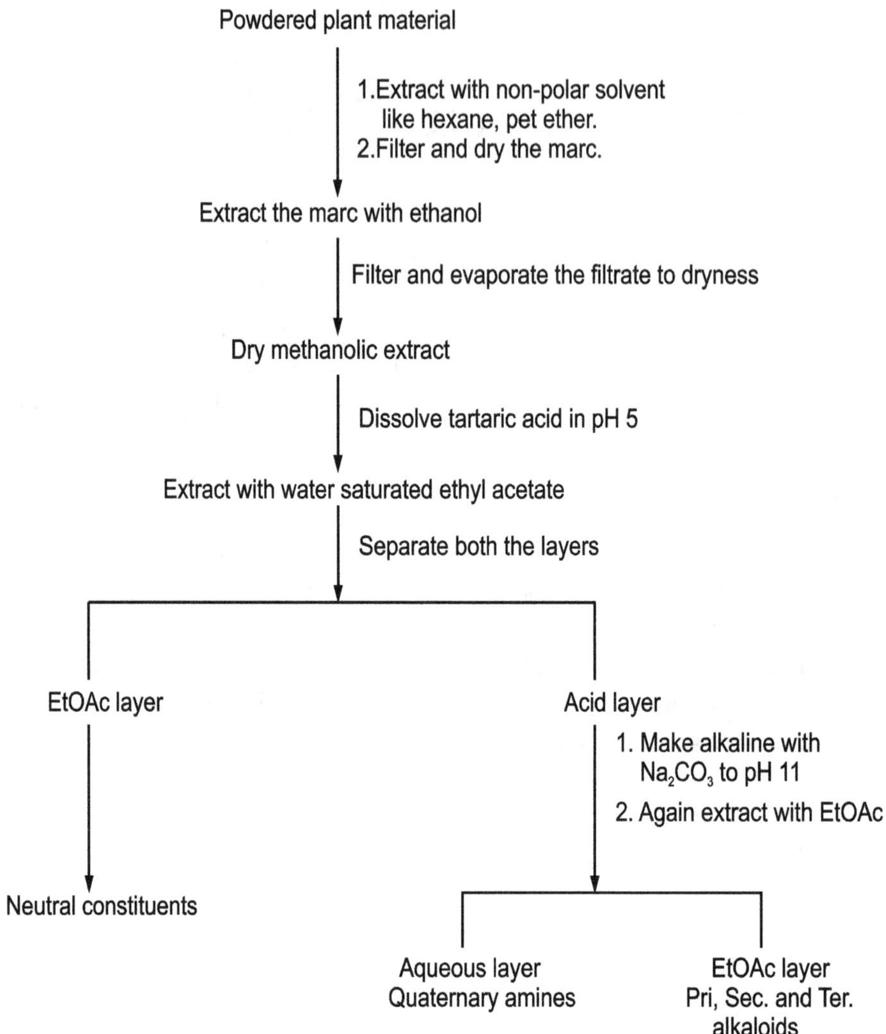

Fig.2.1 General Method for Total Alkaloid Extraction

Kippenberger's Method: In this process, treat powdered plant material with solution of tannin (100 g) in glycerol (500 g) at a constant temperature of 40°C for aduration of 48 hours. Then heat the mixture to 50°C to remove proteins by coagulation, cool to ambient temperature and filter. Shake filtrate with petroleum ether to get rid of fatty materials (oils, fats and waxes). Acidify the fat-free crude plant extract and shake with chloroform, to get total alkaloidal fraction

Stass-otto Method: Moisten the powder of crude drug containing alkaloids with appropriate alkali like ammonia, KOH, Ca(OH)$_2$, NaOH for 1 hour. Extract this alkali treated dried mass with suitable solvent like water-immiscible solvents (e.g., chloroform, ethylene chloride, and benzene or ether) or water-miscible solvents (e.g., methanol and ethanol) which are suitable for extraction of quaternary ammonium alkaloids. Concentrate the organic extract and then add aqueous acid. Separate both the layers and discard organic layer. Treat aqueous layer with any of the selected dilute alkali to liberate the free alkaloid bases. Filter the precipitate and dry the residue to obtain crude alkaloids.

This crude alkaloid can be purified by dissolving the residue in an organic solvent followed by acid-base shake-up or complex formation with a suitable precipitant and decomposition to recover the alkaloids. For example, the tannic acid complex is decomposed by treatment with lead hydroxide or lead carbonate; Mayer's complex is decomposed by hydrogen sulfide; Dragendorff's complex is decomposed by sodium; and picric acid complex is decomposed by ammonia. Fractional precipitation or crystallization separates the alkaloids in the form of salts such as oxalates, tartrates, and picrates. Gradient pH extraction method is suitable for separating alkaloids of different basicity like weak, moderate, and strong basic alkaloids by dissolving crude residue in acidic solution of 2 per cent tartaric acid and extracting with organic solvent. Now, gradually increase the pH to 9.0 and extract again with organic solvent. Chromatographic techniques are the most suitable to isolate single alkaloid from complex mixtures. Distillation is the most suitable method for extraction of liquid or volatile alkaloids like nicotine and coniine.

Indole Alkaloids

Indole alkaloid is one of the largest classes of alkaloids containing more than 4100 known different compounds with significant physiological activity.

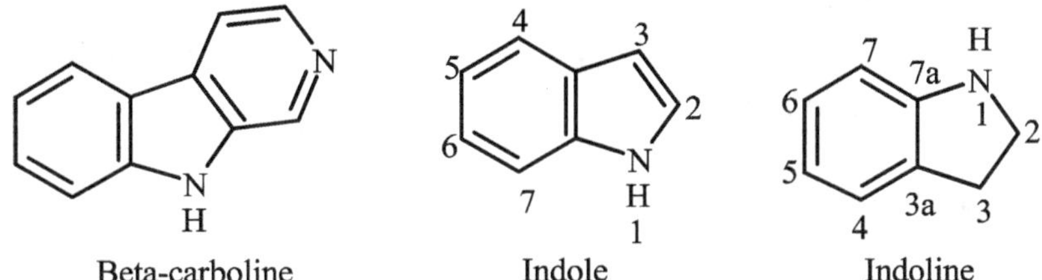

Beta-carboline	Indole	Indoline

These indole alkaloids are biosynthesized from amino acids. Depending on their biosynthesis, two types of indole alkaloids are distinguished; isoprenoids and non-isoprenoids.

Non-isoprenoid	➤ *Simple indole alkaloids*: These are obtained from 5-hydroxy-l-tryptophan. Serotonin is a monoamine. It is a bioactive alkaloid known as a neurotransmitter. It has been found in the cardiovascular system, in blood cells and the peripheral and CNS. Psilocin and psilocybin are the main alkaloids in hallucinogenic mushrooms belonging to the genus Psilocybe. ➤ *β-Carboline alkaloids*: This class includes elaeagnine, harman and harmaline. ➤ *Terpenoid indole alkaloids*: This group of alkaloids contains three types of nucleus

Contd...

	Corynanthe alkaloids: Ajmalicine 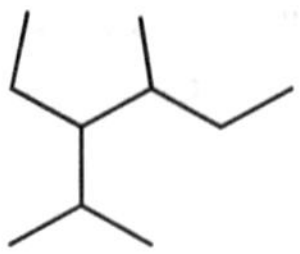 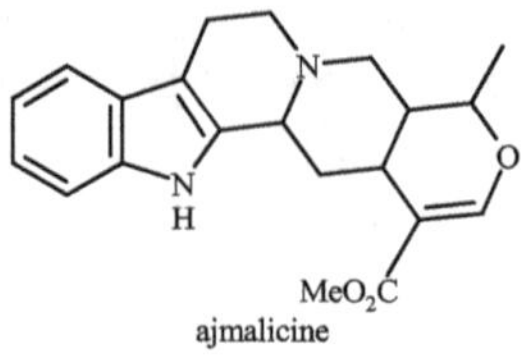 ajmalicine Iboga alkaloids: ibogamine, ibogaine 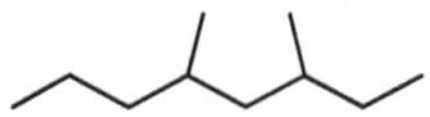 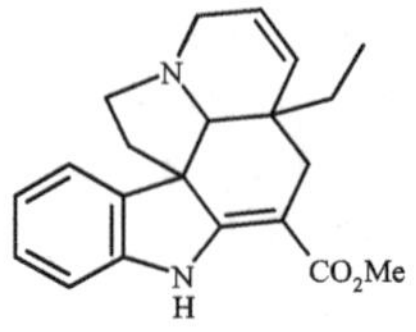 tabersonine Aspidosperma alkaloids: Catharantine 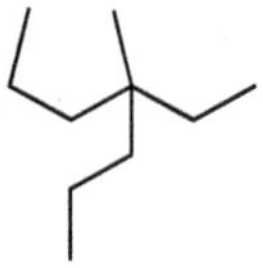 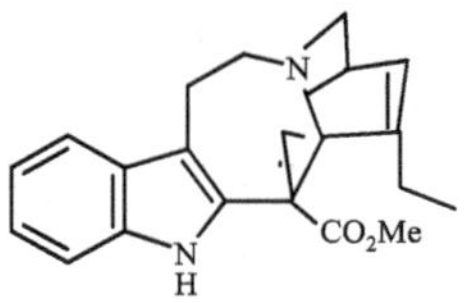 Catharanthine ➢ *Pyrroloindole alkaloids*: The best-known alkaloids belonging to this group are eserine, chimonanthine, eseramine, physo- venine, rivastigmine, eptastigmine, neostig-mine, pyridostigmine and distigmine
Isoprenoid	➢ Hemiterpenoids (Example- ergot alkaloids) and Monoterpenoids.

Isoquinoline Alkaloids

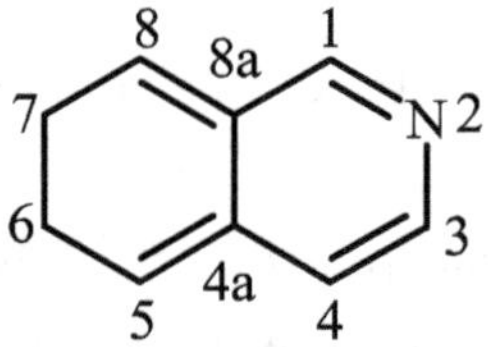

Isoquinoline

They are formed from a precursor of 3,4-dihydroxytyramine (dopamine) linked to an aldehyde or ketone. The Isoquinoline alkaloids are a large class of medicinally active alkaloids whose properties are variable. Their properties include being antispasmodic, antimicrobial, antitumour, antifungal, anti-inflammatory, cholagogue, hepatoprotective, antiviral, amoebicidal, anti-oxidant and can act as enzyme inhibitors. This class notably includes morphine and codeine. They are typically found in the Papaveraceae, Berberidaceae and Ranunculaceae families. They are derived from the amino acids phenylal-anine or tyrosine.

Tropane Alkaloids

Tropane is a bicyclic amine, characterized by a two-ringed structure with a pyrrolidine and a piperidine ring sharing a single nitrogen atom and two carbon atoms. Thus, the common structural element of the tropane alkaloids is the bicyclic azabicyclo-octane skeleton. They are ester alkaloids resulted from the coupling of organic acids with amino alcohol (Base). Main Alkaloids from this class are atropine, hyoscyamine, cocaine, tropinone, tropine, littorine and cuscohygrine. Examples are Atropa, Datura, Hyoscyamus, Duboisia spp. Tropane Alkaloids are classified into:

1. Solanaceous Tropane Alkaloids

2. Erythroxylon (Coca) Alkaloids

1. ***Atropine:*** Chemically it is racemic mixture of hyoscyamine. Its pharma-cological effects are due to binding to muscarinic acetylcholine receptors. It is an antimuscarinic agent. Working as a nonselective muscarinic acetylcholinergic antagonist, atropine increases firing of the sinoatrial node (SA) and conduction through the atrioventricular node (AV) of the heart, opposes the actions of the vagus nerve, blocks acetylcholine receptor sites, and decreases bronchial secretions. Atropine is given as a treatment for SLUDGE (salivation, lacrimation, urination, diaphoresis, gastrointestinal motility, emesis) sym-ptoms caused by organophosphate poisoning. Another mnemonic is DUMBBELSS, which stands for diarrhea, urination, miosis, bradycardia, broncho-constriction, excitation (as of muscle in the form of fasciculations and CNS), lacrimation, salivation, and sweating (only sympathetic innervation using Muscular receptors).

2. ***Hyoscyamine:*** Chemically it is ester of tropine with tropic acid. Hyoscyamine is used to provide symptomatic relief to various gastrointestinal disorders including spasms, peptic ulcers, irritable bowel syndrome, diverticulitis, pancrea-titis, colic and cystitis. It has also been used to relieve some heart problems, control some of the symptoms of Parkinson's disease, as well as for control of respiratory secretions. Racemic form of hyoscyamine is known as Atropine.

3. ***Hyoscine*** **(Scopolamine):** Chemically it is ester of scopine with tropic acid. It is also known as levo-duboisine found in such as henbane, jimson weed (Datura), Angel's Trumpets (Brugmansia), and corkwood (Duboisia). Its primary use is for the treatment of post-operative nausea and vomiting, sea-sickness, leading to use by scuba divers. Scopol-amine is used as an adjunct anesthetic in trauma surgery. Light anesthesia is required to keep the patient hemo-dynamically stable, and this increases risk of awareness. Scopolamine decreases awareness and recall. It also used as antispasmodic. Racemic form of hyoscine is known as Atroscine.

(-) Hyoscine
(scopolamine)

Hyoscyamine
(Daturine)

Atropine
(Racemic mixture of (-)
hyoscyamine)

Cocaine

Different tropane alkaloids

Rauwolfia

Synonym: Rauwolfia root, Serpentina root, Hindi-Chhotachand, Marathi-Serpagangha, candrika,chandramarah

Biological source: It is dried roots of plant *Rauwolfia serpentina* (*Ophioxylonserpentinum*) of Family Apocynaceae.

Geographical source: Asia, America, Africa, India, Sri lanka, Mynmar, Thialana.

Morphology: Colour - reyish yellow to brown, Odour: Odourless, Taste: Bitter, Size: 10-18 cm long and 1-3 cm in diameter, Shape: sub-cylindrical, slightly tapering, tortuous; Extra features: longitu-dinal wrinkled surface. Few circular root scars with tetrastichous arrangements, short fracture

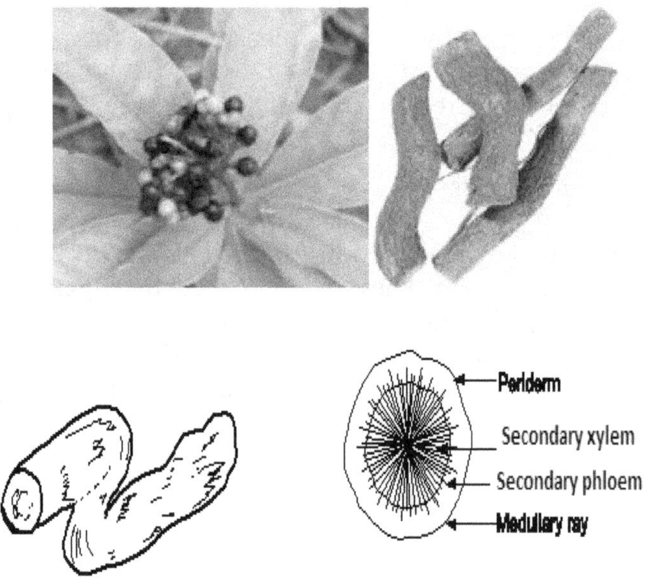

Fig.2.2

Microscopy: Transverse section shows reddish-brown cork composed of three or four layers of thin walled, isodiametric cells; lignified parenchyma consisting starch granules from medullary rays, border pitted xylem vessels, scattered prismatic calcium oxalate crystals in parenchymatous cells, parenchyma cells from the phello-derm and phloem consisting starch grains, unlignified pericyclic fibres and tetrarch xylems.

Chemical constituents: Three types of alkaloids are present in *Rauwolfia serpentina*.

- Weakly basic Indole Alkaloids: tertiary Indole Alkaloids like Reserpine, Rescinnamine, despiridine

- *Intermediate basic Indoline* (2, 3, *dihydro structure*) *Alkaloids*: tertiary Indoline alkaloids Reserpiline, Ajmaline, Iso- Ajmaline, rauwolfinine.

- *Strong Basic alkaloides*: Serpentine, serpentinine and alsotonine are strongly basic anhydronium alkaloids

- While Ajmalinine, Ajmalicine, Chandrine, renoxidine, reserpinine, Sarpagine, Tetraphyllicine, Yohimbine, 3-epi-ayohimbine are the other alkaloids present in Rauwolfia

Ajmaline was first isolated by Salimuzzaman Siddiqui in1931fromthe roots of *Rauwolfia serpentine* and he named it ajmaline, after Hakim Ajmal Khan, one of the most illustrious practitioners of Unani medicine in South Asia

rescinnamine

yohimbine

serpentine

Reserpine

Chemical Tests

Reserpine:Treat the reserpine powder with vanillin -acetic acid solution. Violet red colour.

Uses: Snakeroot depletes catecholamines and serotonin from nerves in central nervous system. Reserpine is used as antipsychotic and antihypertensive drug that has been used for the control of high blood pressure and for the relief of psychotic symptoms as it blocks vesicular monoamaine transporter (VMAT).

Reserpine and similar alkaloids are efficacious in reducing arterial pressure when used with a diuretic. Reserpine is generally added to the treatment regimen if response to a thiazide (or thiazide-like) diuretic is inadequate. Reserpine is also useful in treating hypertensive emergencies. The maintenance dose of reserpine is up to 0.1-0.25 mg daily. Because reserpine has a long half-life, a loading dose is employed to obtain a reasonably rapid steady-state concentration.

Reserpine is contraindicated absolutely in depression and in those with a history of depression. The drug is also contraindicated in severe renal failure and is best avoided in peptic ulceration, ulcerative colitis, or asthma. Reserpine may cause complications in the neonate if used in pregnancy. These include nasal obstruction (anosmia), bradycardia, and hypothermia. Thus, reserpine is no longer a drug of choice in hypertensive emergencies in pregnancy. A reduced dose is recommended in the elderly.

Several drug interactions have been reported. There is enhanced peripheral vasodilatation and hypotension with alcohol. Enhanced falls in blood pressure are also seen with glyceryl trinitrate, L-dopa, fenfluramine, and phenothiazines. The pressor effects of phenylephrine and catecholamines are enhanced, whereas the effect of direct-acting amines (such as ephedrine) is diminished. There is excessive central nervous system excitation with monomine oxidase inhibitors.

Yohimbine is useful in treatment of erectile dysfunction. Ajmaline acts as anti-arrythmic agent. Ajmalicine, also known as δ-yohimbine or raubasine, exerts antihypertensive action. Recinnamine acts as angiotensin converting enzyme inhibitor. Serpentine has mild sedative action.

Vinca

Synonym: Catharanthus, Periwinkle.

Biological source: It is dried whole plants of *Catharathus roseus* of Family Apocynaceae.

Geographical source: Indigenous to Madagascar, cultivated in South Africa, India, U.S.A., Europe, Australia and Caribbean island.

Morphology: Color- leaves are green, flowers are pink, white or red; odour-disagreeable, taste-bitter, fruits are many seeded

Microscopy: Transverse section shows a dorsiventral leaf having upper epidermis, lower epidermis and single layer of palisade cells just below the upper epidermis, upper epidermis composed of cuticulaised rectangular thin walled cells and less number of anisocyticstomatas and covering trichomes, elongated and compact palisade cells, thick walled cellulosic collenchymas, lignified xylem vessels and cortical parenchyma. Calcium oxalate crystals are absent.

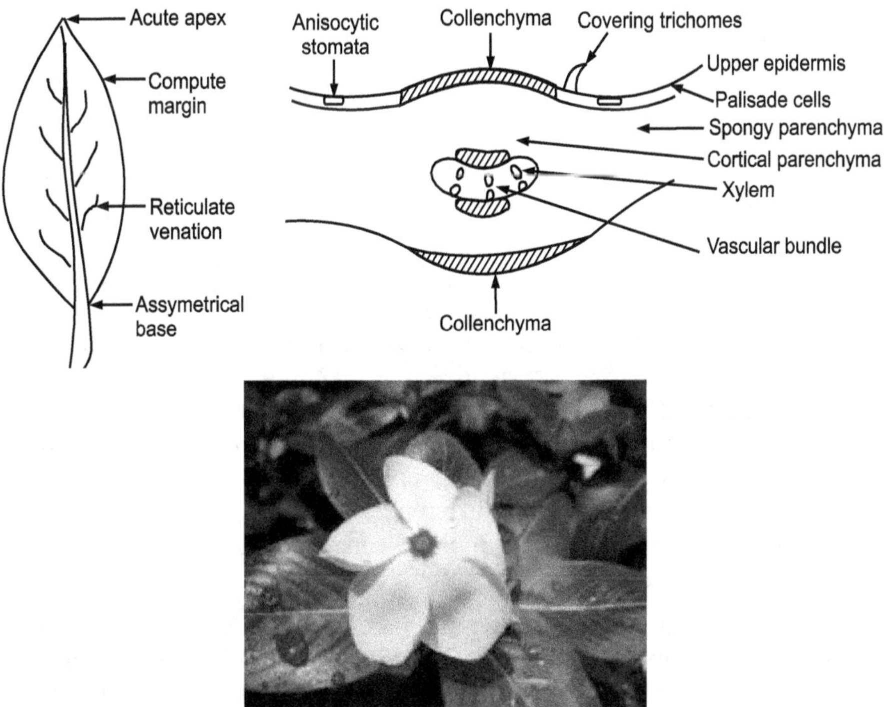

Chemical constituents: Dimeric dihydro-indol alkaloid vinblastin, vincrystine, vindoline, ajmalicine, lochnerine, serpent-tine, tetrahydroalstenine. Catharanthine is one of the two precursors that form vinblastine, the other being vindoline.

Contd...

<table>
<tr><td style="text-align:center">Catharanthine</td><td style="text-align:center">Vincamine</td></tr>
</table>

Uses: Tubulin is a structural protein that polymerizes to microtubules. The cell cytoskeleton and mitotic spindle, among other things, are made of microtubules. Vincristine binds to tubulin dimers, inhibiting assembly of microtubule structures. Disruption of the microtubules arrests mitosis in metaphase. Therefore, the vinca alkaloids affect all rapidly dividing cell types including cancer cells, but also those of intestinal epithelium and bone marrow. Vincamine acts as a vasodialator.This synthetic structural differences influence their activity. Vinblastine is used to treat Hodgkin's disease (a form of lymphoid cancer), while vincristine is used clinically in the treatment of children's leukaemia. Vincristine is more neurotoxic than vinblastine.

Vinblastine is indicated in the treatment of patients with Hodgkin's and non-Hodgkin's lymphomas, breast cancer, Kaposi's sarcoma, renal cell cancer, and testicular cancer. Vincristine is more widely indicated, including for the treatment of patients with myeloma, acute lymphocytic leukemia, Hodgkin's and non-Hodgkin's lymphomas, rhabdomyosarcoma, neuroblastoma, Ewing's sarcoma, Wilm's tumor, chronic leukemia, thyroid cancer, brain tumors, and trophoblastic neoplasia. Adverse reactions are broad, including typical side effects of cytotoxic chemotherapeutics, such as myelosuppression, mucositis, fever, anemia, and alopecia. Vincristine also causes additional side effects, such as hypertension, neuropathy, depression, Raynaud's phenomenon, myocardial infarction, and pulmonary edema. Resistance mechanisms include gp170-mediated MDR and mutations in tubulin subunit proteins that decrease drug binding.

Vinorelbine is a semisynthetic derivative of vinblastine that also inhibits tubulin polymerization and disrupts spindle assembly in the M phase. This compound has a higher specificity for mitotic microtubules and a lower affinity for axonal microtubules, reducing neuropathy. Vinorelbine is indicated in the treatment of lung cancer, breast cancer, and ovarian cancer.

Opium

Synonym: Raw opium

Biological source: It is dried latex obtained by incision from the unripe capsule of *Papaverumsomniferum* family Papaveraceae.

Geographical source: India, Pakistan, Afghanistan, Turkey, Russia, China, and Iran.

Morphology: Colour: Dark brown; Taste: Bitter; Odour: Strong characteristic; Shape: Cubical (Indian), Brick (Turkish), Oval (manipulated Turkish), elongated (mani-pulated European)

Chemical constituents: Morphine was first isolated between 1803 and 1805 by German pharmacist Friedrich Sertürner. This is generally believed to be the first isolation of an active ingredient from a plant. Merck began marketing it commercially in 1827. Morphine was more widely used after the invention of the hypodermic syringe in 1853–1855. Sertürner originally named the substance morphium, after the Greek god of dreams, Morpheus, as it has a tendency to cause sleep.Opium contains approximately 12% morphine, The latex also includes codeine and non-narcotic alkaloids such as oripavine, papaverine, thebaine and noscapine. Morphine is frequently processed chemically to produce heroin for the illegal drug trade.

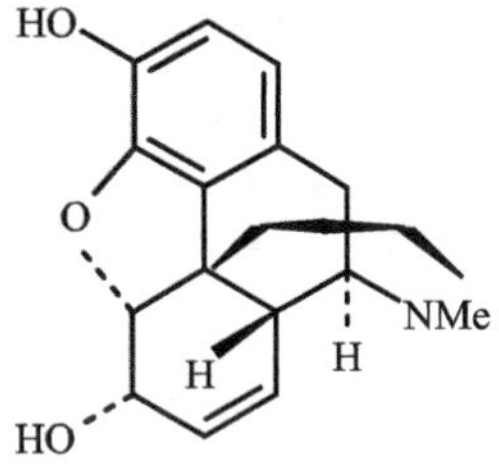

Morphine

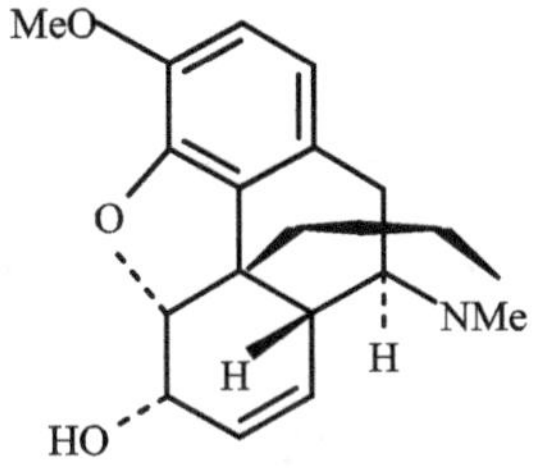

Codeine

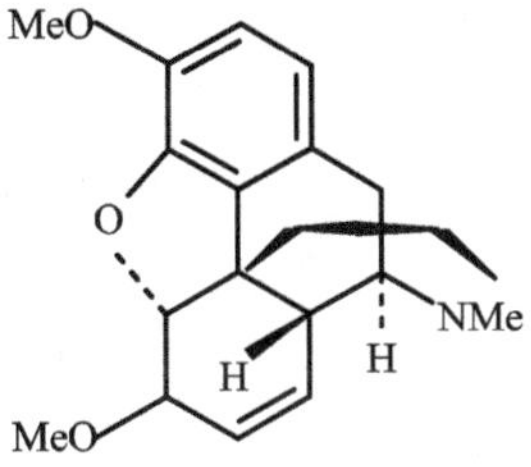

Thebaine

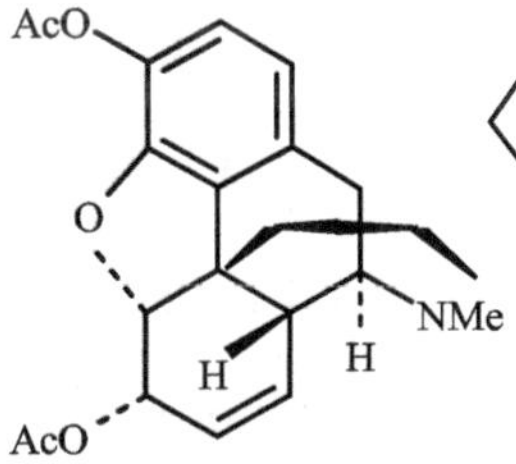

Heroin
(diamorphine)

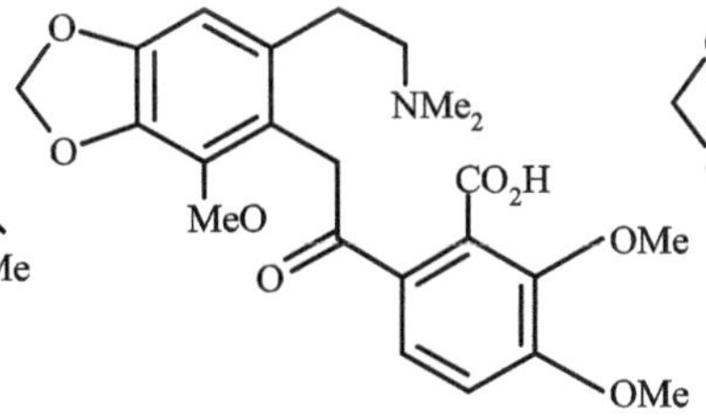

Narceine

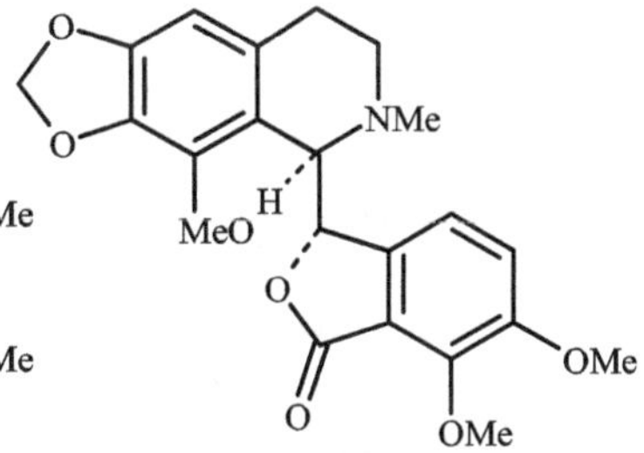

Noscapine
(Narcotine)

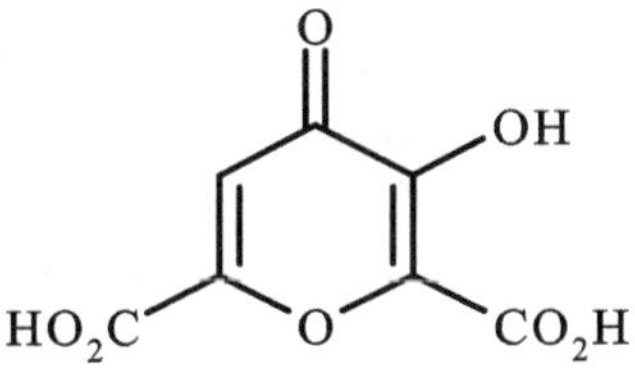

Meconic acid

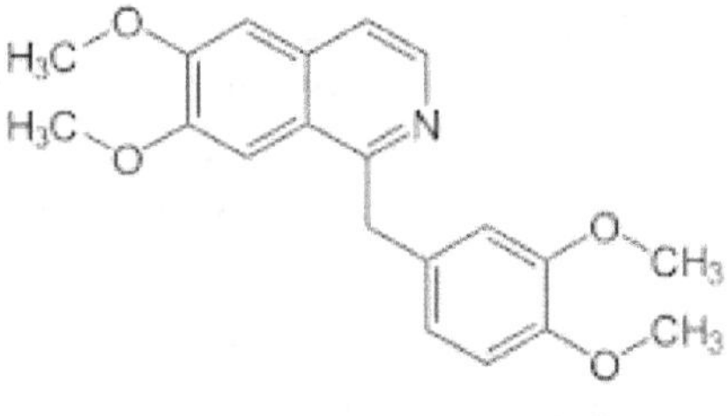

Chemical Tests

Morphine HNO$_3$ test	Treat the morphine powder with nitric acid	Orange red colour.
Morphine H$_2$SO$_4$ test	Treat morphine with concentrated sulphuric acid and formaldehyde	Dark violet colour.
Papaverine	Treat papaverine solution with HCl and potassium ferricyanide solution	Lemon yellow colour.

Analgesic Efficacy	Origin	Function

Strong
  Morphine
  Pethidine
  Fentanyl
  Alfentanil
  Remifentanil
  Sufentanil
 	
Intermediate
  Buprenorphine
  Pentazocine
  Butorphanol
  Nalbuphine
 	
Weak
  Codeine
	Naturally occurring
	  Morphine
	  Codeine
	  Papavarine
	  Thebaine
	
	Semisynthetic
	  Diamorphine
	  Dihydrocodeine
	  Buprenorphine
	
	Synthetic
	Phenylpyperidines:
	  pethidine, fentanyl, alfentanil,
	  sufentanil
	Diphenylpropylamines:
	  methadone, dextropropoxyphene
	Morphinans:
	  butorphanol, levorphanol
	Benzomorphans:
	  pentazocine
		Pure agonists
		  Morphine
		  Fentanyl
		  Alfentanil
		  Remifentanil
		  Sufentanil
		
		Partial agonist
		  Buprenorphine
		
		Agonists-antagonists
		  Pentazocine
		  Nalbuphine
		  Nalorphine
		
		Pure Antagonists
		  Naloxone
		  Naltrexone

Fig.2.3 Different opium alkaloids and their semisynthetic/synthetic derivatives

Uses: Opium contains two main groups of alkaloids. Phenanthrenes such as morphine, codeine, and thebaine are the main narcotic constituents. Isoquinolines such as papaverine and noscapine have no significant central nervous system effects,and are not regulated under the Controlled Substances Act.

Morphine is a phenan-threne opioid receptor agonist – its main effect is binding to and activating the delta and mu opioid receptors in the central nervous system which is associated with analgesia, sedation, euphoria, physical dependence, and respiratory depression. Morphine is a rapid-acting narcotic. Processing of pain information is inhibited by a direct spinal effect at the dorsal horn.Probably involves presynaptic inhibition of the release of tachykinins like substance P.

Morphine and narceine are active on µ- and κ- opiate receptors. They are also known as analgesic agents. These alkaloids may be used as pain relievers. Narceine is also known to be used in the treatment of a cough.

Codeine relieves local irritation in the bronchial tract and as an antitussive. It is mild analgesic, relaxant to smooth muscles.

Thebaine has stimulatory rather than depressant effects and at high doses can cause convulsions. Thebaine is not used therapeutically, but it is used as a starting compound in synthesis of compounds like oxycodone, oxymorphone, nalbuphine, naloxone, naltrexone, buprenorphine and etorphine.

Oripavine possesses an analgesic potency comparable to morphine; however, it is not clinically useful due to severe toxicity and low therapeutic index.

The effects of morphine can be countered with opioid antagonists such as naloxone and naltrexone; the development of tolerance to morphine may be inhibited by NMDA antagonists [class of anesthetics that work to antagonize, or inhibit the action of, the N-Methyl-D-aspartate receptor (NMDAR)] such as ketamine or dextromethorphan. Diacetylmorphine better known as heroin is synthesized from morphine able to cross the blood–brain barrier faster than morphine due to the lipid solubility of diacetylmorphine.

Papaverine is a direct smooth muscle relaxant. It is anti-spasmodic and vasodialator.

Belladonna

Synonym: Belladonna leaf, Belladona folium, Deadly night shade leaf.

Biological source: It is dried leaves and aerial part of *Atropa belladonna of* Family Solanaceae.

Geographical source: England, European countries, Asian countries.

History: The species name "belladonna" ("beautiful woman" in Italian) comes from the original use of deadly nightshade to dilate the pupils of the eyes for cosmetic effect. Both atropine and the genus name for deadly nightshade derive from Atropos, one of the three Fates who, according to Greek mythology, chose how a person was to die. It was Atropos who chose the mechanism of death and ended the life of each mortal by cutting their thread with her "abhorred shears."

Morphology: Color: leaves green, flowers purple and fruits green; Odour: slight; Taste: Bitter; Extra feature: crumpled and twisted

Microscopy: TS of belladonna leaf shows presence of upper epidermis, lower epidermis, palisade cells, calcium oxalate crystals, anisocytic stomata, covering as well as glandular trichomes.

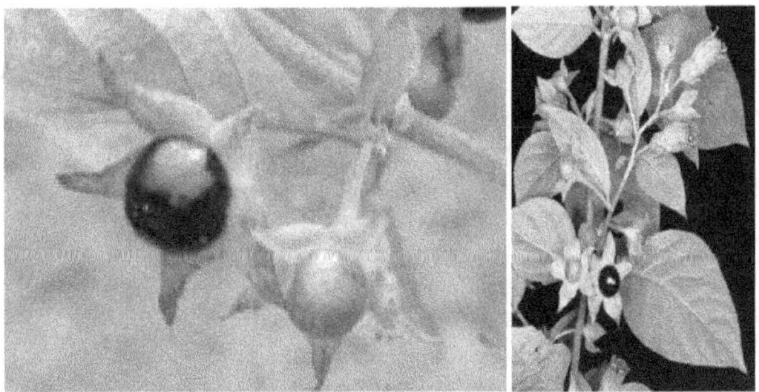

Chemical constituents: It contains tropane alkaloids majorly l-hyoscymine and atropine. It also contains traces of hyoscine, homotropine, belladonnine and scopoletine.

Chemical Tests

Atropine	**Vitali-Morin's test:** Take small quantity of the solid atropine and add 2 drops of conc. nitric acid are mixed in an evaporating dish and evaporated to dryness on water bath to the faintly yellow residue, cool well. Then dissolve the residue in 1 ml of acetone, add few drops of freshly prepared solution of alcoholic potassium hydroxide.	A color (bright purple) is formed which changes to color (red) and fades gradually to cc *Contd...*
Atropine	**Gerrard reaction:** Take 1 ml. of 2% solution of mercuric chloride (in 50% alcohol) and add few mg of atropine.	A red color will be produced

Uses: Atropine and l-hyosciamine are antimuscarinic agents used to reduce the secretion such as sweat, saliva and gastric juice. It is effective as adjunctive therapy in the treatment of irritable bowel syndrome (irritable colon, spastic colon, mucous colitis) and acute enterocolitis.

The symptoms of belladonna poisoning include dilated pupils, sensitivity to light, blurred vision, tachycardia, loss of balance, staggering, headache, rash, flushing, severely dry mouth and throat, slurred speech, urinary retention, constipation, confusion, hallucinations, delirium, and convulsions.The antidote for belladonna poisoning is an anticholinesterase (such as physostigmine) or a cholinomimetic (such as pilocarpine), the same as for atropine.

2.2   Phenylpropanoids and Flavonoids

Phenylpropanoids

The phenylpropanoids are compounds which contain aromatic phenyl ring and three carbon chain (propene tail). Most of them are derived from amino acids phenylalanine and tyrosine. These compounds protect plants from infections, wounding, UV irradiation, exposure toozone, pollutants, and herbivores. Following are different types of phenylpropanoids:

➢ Hydroxycinnamic acids derived: coumaric acid, caffeic acid, ferulic acid, 5-hydroxyferulic acid, and sinapic acid

➢ Cinnamic aldehydes and monolignols derived: cinnamaldehyde, coumaryl alcohol, coniferyl alcohol, and sinapyl alcohol

➢ Coumarins and flavonoids derived: umbelliferone,4-coumaroyl-CoA, chalcones

➢ Stilbene and Stilbenoids derived: resveratrol

Phenylalanine

Tyrosine

Cinnamic acid

Coniferyl alcohol

Safrole

Umbelliferone

Chalcones

Flavonoids

Stilbene

Lignans and Lignin

Lignins, known as plant polymer, are obtained from phenolic oxidative coupling of hydroxycinnamyl alcohol monomers, brought about by peroxidase enzymes. Para coumaryl alcohol, coniferyl alcohol and synapyl alcohol are three major building blocks of lignin. he term lignan is used specifically to molecules in which the two phenylpropane units are coupled at the central carbon of the side-chain like in pinoresinol, whilst compounds containing other types of coupling as in guaiacylglycerol β-coniferyl ether and dehydrodiconiferyl alcohol, are then referred to as neolignans. Lignan/neolignan are immense storage area of aromatic compounds but difficult to release these metabolites.

p- coumaryl alcohol

coniferyl alcohol

synapyl alcohol

Flavonoids

Over 5000 naturally occurring flavonoids have been characterized from various plants. Generally, flavonoids are crystalline solids, and only a few of them are amorphous powders. Flavanones and flavanonols are colorless. Chalcones are yellow-orange color. Flavones, flavonols and their glycosides are yellow in color. Color of anthochyanidines are varies with pH ranges i.e. pH < 7 = red; pH = 8.5 = purple; pH > 8.5 = blue. All flavonoid glycosides are optically active due to presence of sugars. Flavonoid glycosides are soluble in water and methanol.

Extraction: Take accurately weighed dried material. If flavonoid aglycones like isoflavones, flavonols, flavanones are present, then extract with non-polar or less polar solvents (Example: chloroform, diethyl ether or dichloromethane). If flavonoid glycosides are present then extract with polar solvents like alcohols or aqueous alcohols. Fractionation of alcoholic extract with ethyl acetate separates most of semi-polar to polar flavonoid from mixtures but non-polar flavonoid need to be separated by column chromatography. Anthocyanidines need to be extracted with acidic water or actone.

Flavnoid classes and examples

Class	Example
	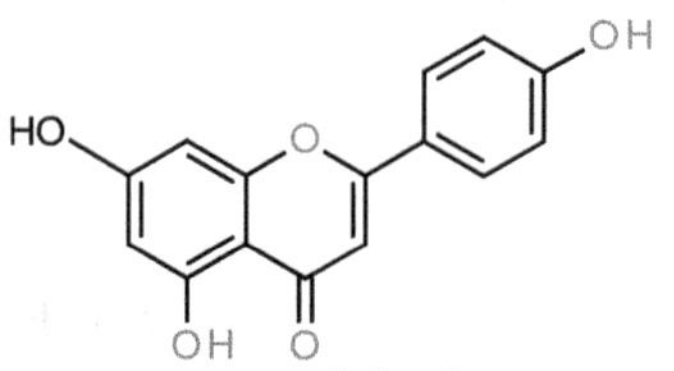 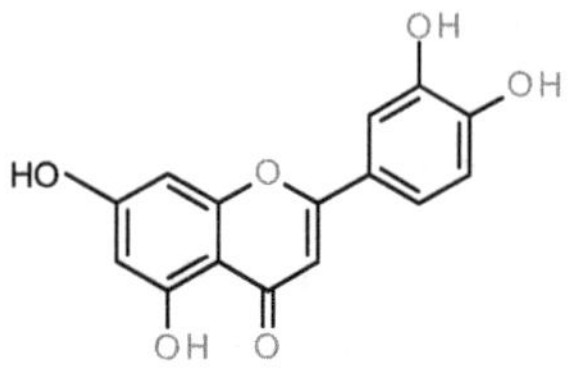 Apigenin Ludolonin
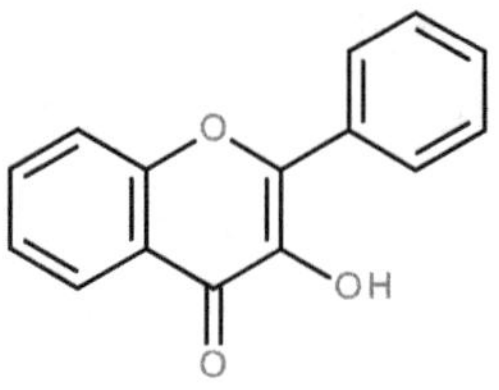 Flavanol	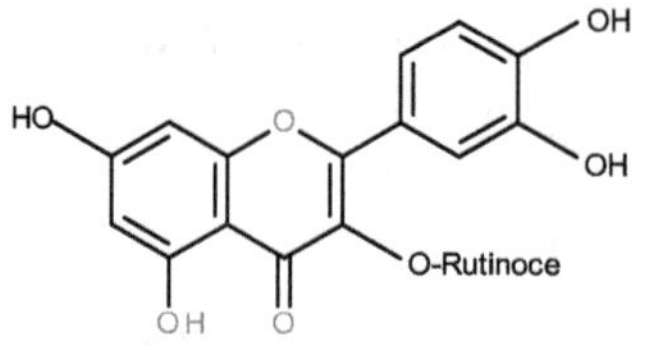 Rutin (sophorin) 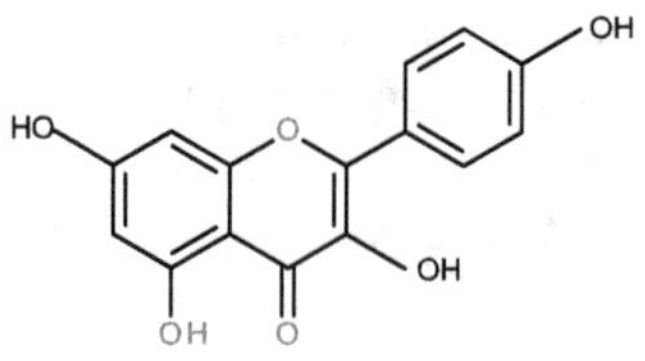 Kaamphenol 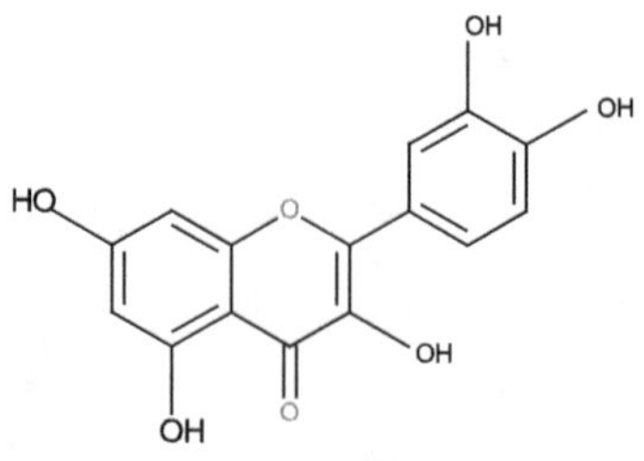 Quercetin

Contd…

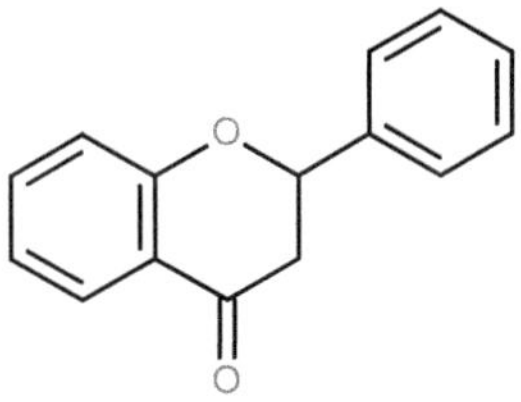 Flavanone	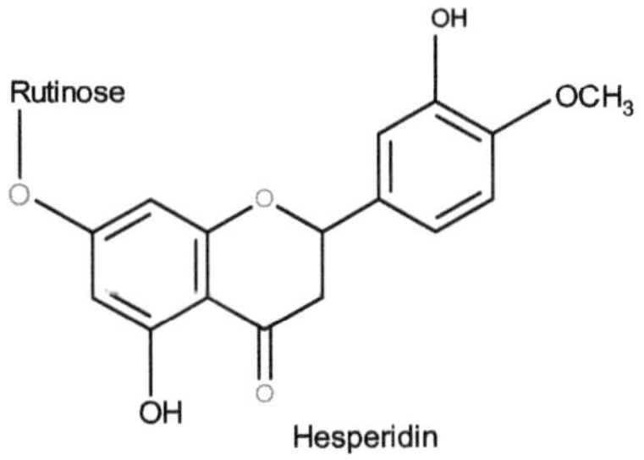 Hesperidin 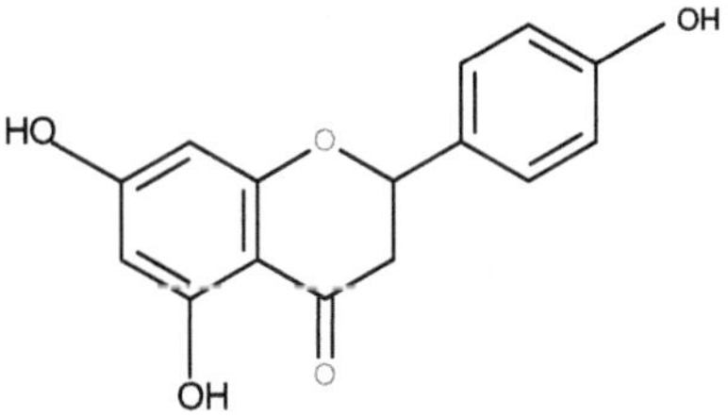 Naringenin
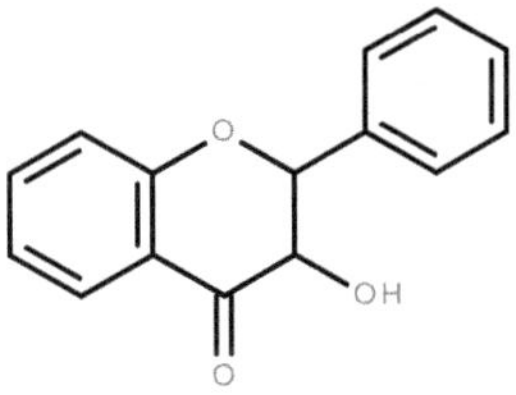 Flavanoids	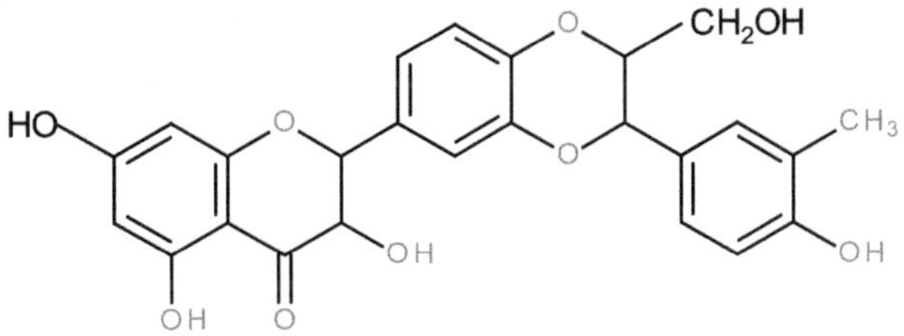 Silibinin
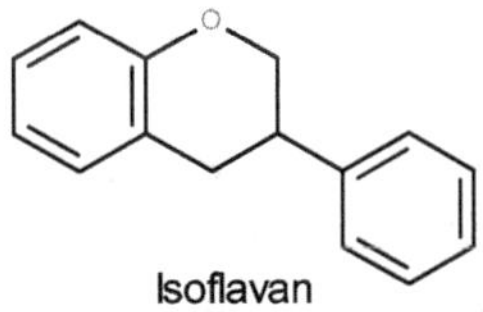 Isoflavan	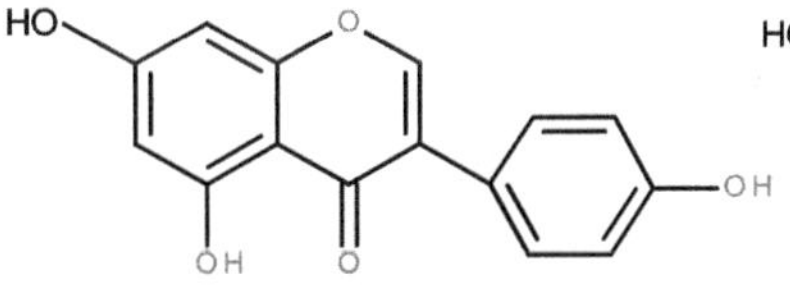 Dubbein 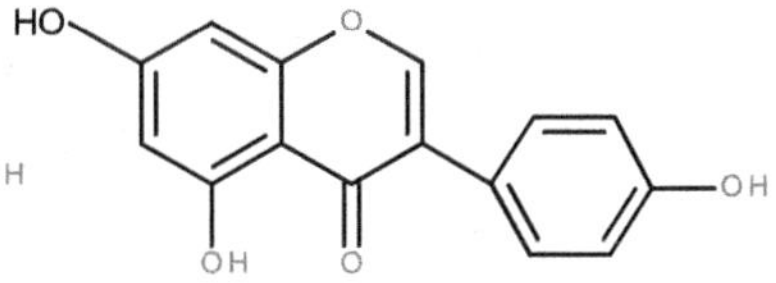 Genbedin
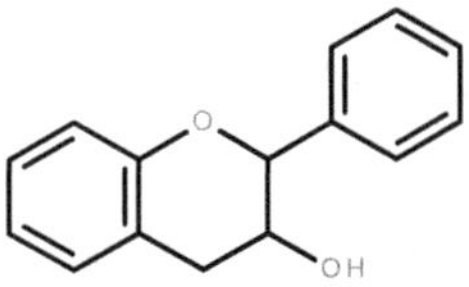 Flavan-3-oil	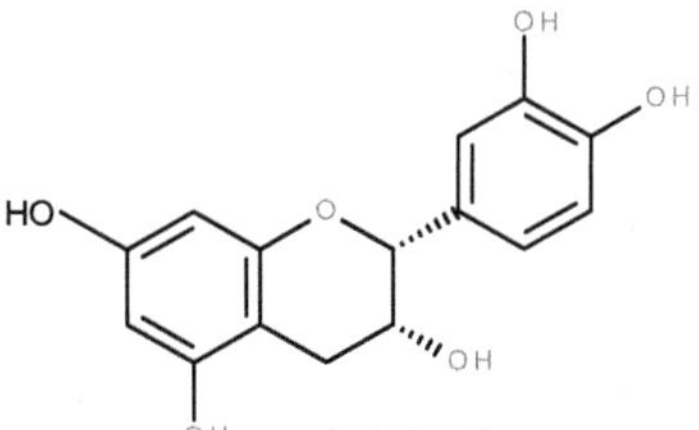 Epicatechin 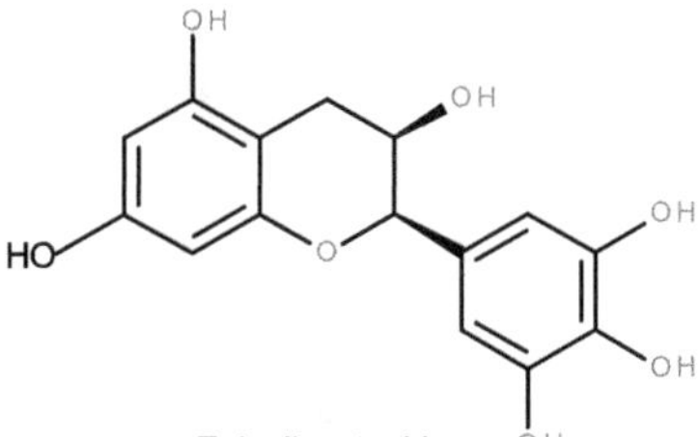 Epigallocatechin
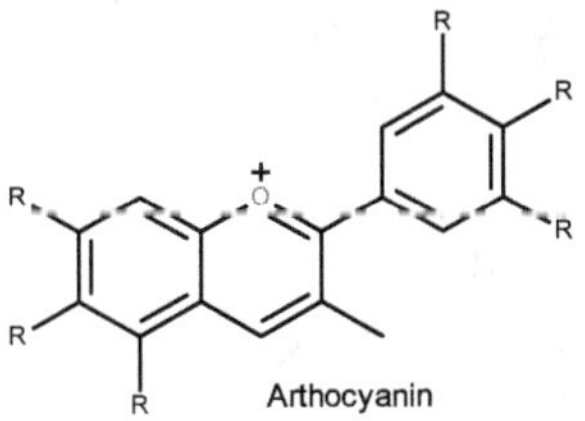 Arthocyanin	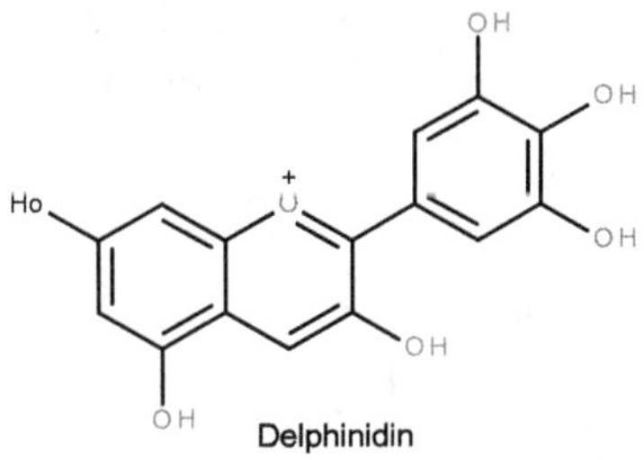 Delphinidin 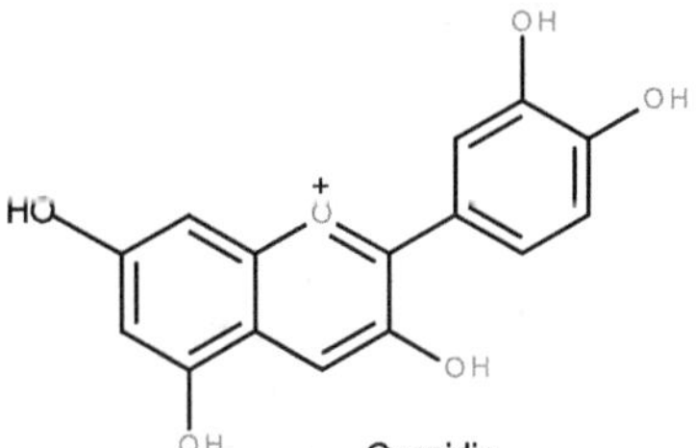 Cyanidin Note: Anthocyanins (colored pigments) are glycosides of anthocyanidins. Anthocyanidins are the sugar-free counterparts of anthocyanins.

Contd...

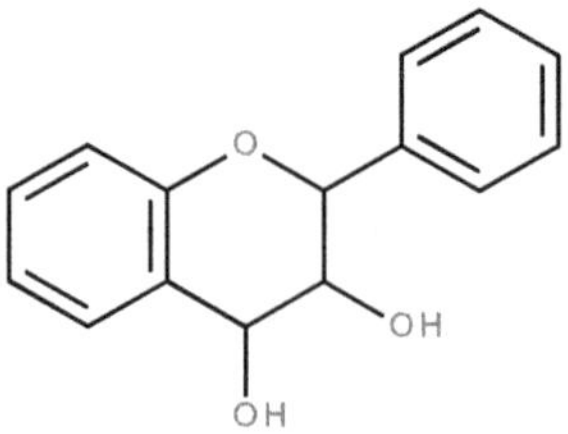 Leucoanthocyanin (flavan-3,4-diols)	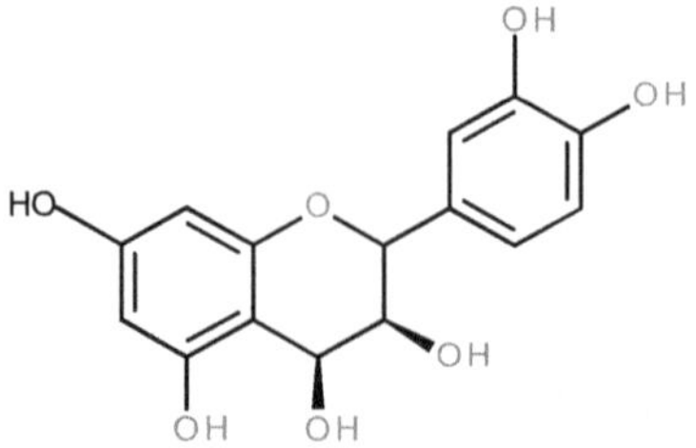 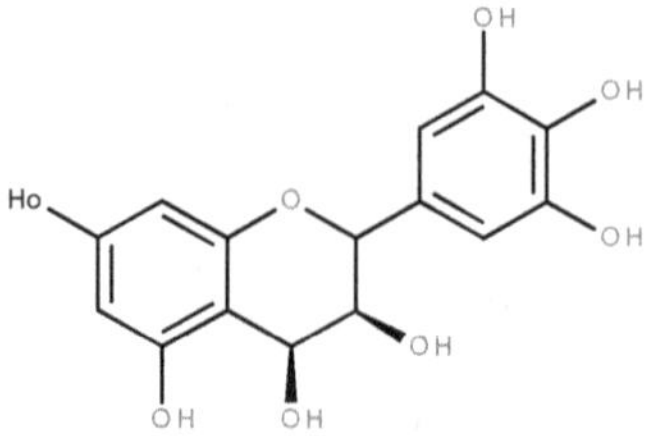 Leucocyanidin Leucodelphinidin Note:Leucoanthocyanin are colorless chemical compounds

Fig.2.4 Various flavonoid drugs

Tea

Synonym: Camellia thea, Chaha, Chai

Biological source: It is dried leaves and buds of *Thea sinensis* of family Theaceae. Tea is the second most consumed beverage on Earth after water.

Geographical source: China, Japan, India, Srilanka

Preparation: The first crop of leaves is gathered in 1-3 years after planting. Only the stiff ripe leaves, easily detached, are collected. Without careful moisture and temperature control during manufacture and packaging, the tea may become unfit for consumption, due to the growth of undesired moulds and bacteria. At minimum, it will make the taste unpleasant. Black tea is prepared by allowing enzymatic oxidation through fermentation of leaves chrorophyllbraks down and polyphenols releases. Green tea is prepared by steaming and drying but without allowing oxidation of leaves.

Morphogy: Leaves color: dark green, Odour: characteristics, Taste: Bitter, Blunt at apex and tapering base

Chemical constituents: Tea leaves contains mainly tannins (catechin) and purine alkaloids (caffeine, theophylline, theo-bromine). L-theanine is amino acid present in tea leaves responsible for modulation of caffeine's psychoactive effect and contributes tea's umami taste. Tannins and volatile oil (geraniol, linalool, benzyl alcohol) are also responsible for flavor of tea.

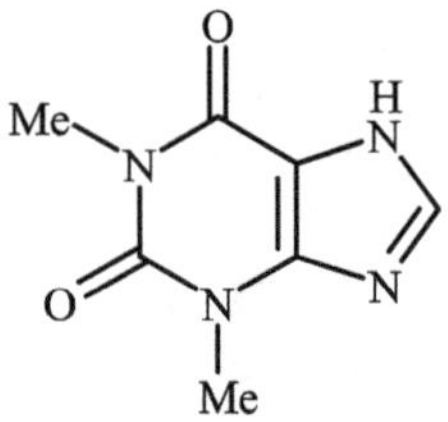

Theophylline
(1, 3-dimethylxanthine)

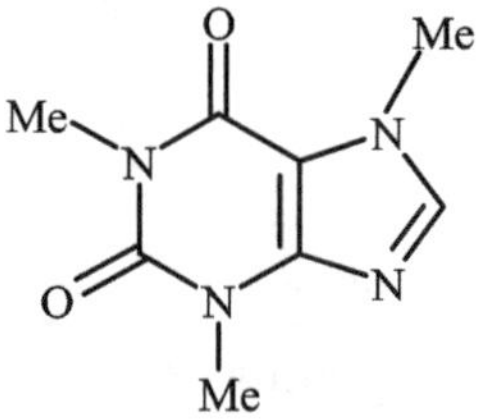

Caffeine
(1, 3, 7-trimethylxanthine)

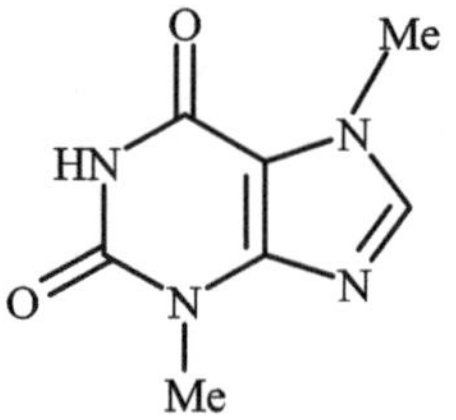

Theobromine
(3, 7-dimethylxanthine)

Chemical Tests

Murexide test for purines	Take few gram of caffeine and add 2-3 ml bromine water. Evaporate to dryness. Then add 2-3 drops of ammonia solution	Purple coloration

Uses: Caffeine is central nervous system stimulant. Mode of action is adenosine recptor antagonist which promotes neurotransmitter release and phopshodieasterase enzyme inhibition which increases cyclic AMP. Theobromine and theophylline are muscle relaxant and diuretic. Alkaloids are also useful in treatment of asthma, chronic pulmonary disease (COPD) and respiratory disease. Thea flavines are potent antioxidant principles which are respon-sible for prevention of cancer and many other diseases.

Ruta

Common name: Rue, Sadab or Satab

Biological source: It is a strongly odoriferous evergreen herbs or a small shrubRuta graveolens L. belonging to the family Rutaceae.

Morphology: It is a perennial, scented and glabrous herb or a sub-shrub. Stem is slender, smooth, pale glaucous green and reaches up to a meter in height. Leaves are alternate, gland-dotted, glaucous, compound, 2-3 pinnate. Leaf-lets are linear-oval or oblong. Inflorescence is terminal corymbose, irregularly dichotomous cymes. Flowers are regular bisexual, terminal ones are pentamerous and others are tetramerous. Petals are distinct, widely spreading, greenish yellow, wide and hooded at top, abruptly connected to narrow claw below, margin wavy and sometimes toothed. Fruits are dry, hard, roundish, 4-5 blunted lobed at top. Color: green. Odor: aromatic and pleasant, Taste:

Microscopy: Typical dicot structure with single layered epidermis, double layered palisade cells and loosely arranged spongy cells. Few Rosette type of calcium oxalates are present and trichomes absent.

Chemical constituents: More than 120 natural compounds mainly including acridone alkaloids, coumarins, essential oils, flavonoides, and fluoroquinolones have been found in the roots and aerial parts. Rutin, quercetin, psoralen, methoxypsoralen, rutacridone, rutacridone epoxide and gravacridondiol are phytochemical compounds which are reported from this plant. α-Pinene, limonene and 1,8-cineole were identified as the main monoterpene constituents for R. graveolensessential oil.

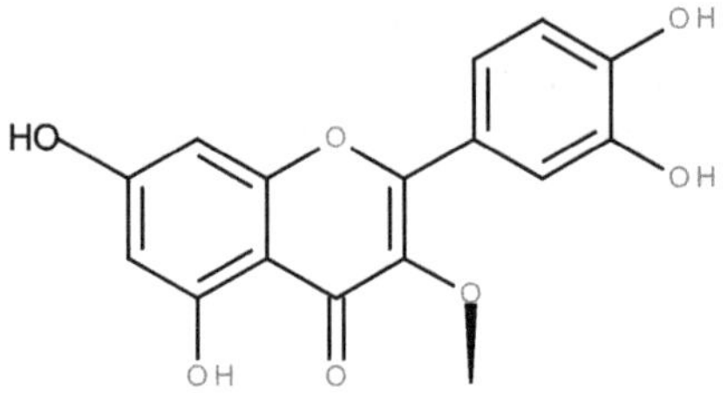

Uses: Rue imparts a pungent aroma, it is combined with other botanicals and used in the production of fragrances and cosmetic toiletries. In traditional system of medicine it is used to srelief of pain, eye problems, rheumatism and dermatitis. Stimulant, emmenagogue, diuretic, abortefacient, and resolvent.

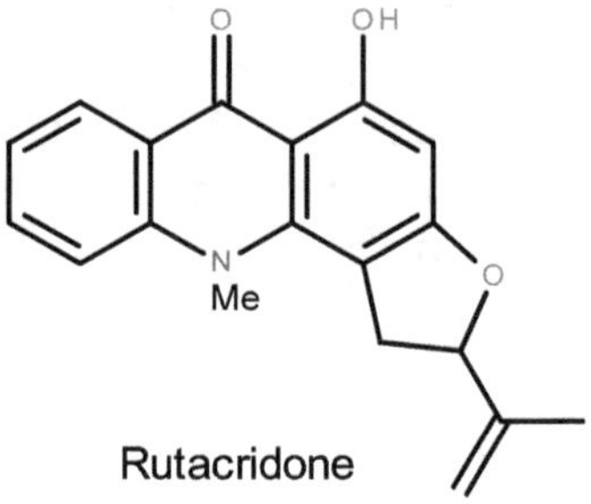

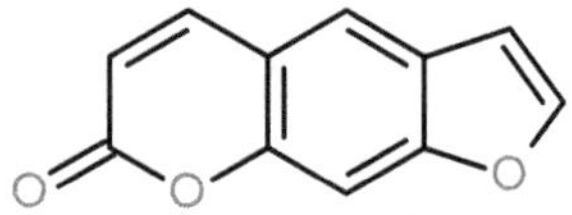

2.3   Steroids, Cardiac Glycosides and Triterpenoids

2.3.1   Steroids (Steroidial Saponin Glycosides)

Liquorice

Synonyms: Glycyrrhiza, Liquorice root, Gly-cyrrhizae radix, Mulethi

Biological Source: It is dried, peeled or unpeeled, root and stolons of *Glycyrrhiza glabra* Linn., belonging to family Leguminosae.

Geographical Source: The drug is commercially cultivated on a large scale in Spain, Sicily and England. *Glycyrrhiza glabra* var. *giandulifera* (Russian liquorice) grows in Russia and Glycyrrhiza glabra var. viotacea comes from Iran.

Morphology: Colour: yellowish-brown or dark brown externally, Odour: Faint and characteristic, Taste: Sweet, Size: Length 20 to 50 cm and diameter 2 cm, Shape: Cylindrical pieces which are straight may be peeled or unpeeled. Peeled liquorice is angular, Fracture: fibrous and splintery.

Microscopy: Transverse section shows orange-brown cork composed of thin-walled polygonal cells, rounded to rectangular thin-walled parenchyma from the cortex, medullary rays and pith, lignified bordered pitted vessels, groups of fibers surrounded by prisms of calcium oxalate crystals, small spherical starch granules.

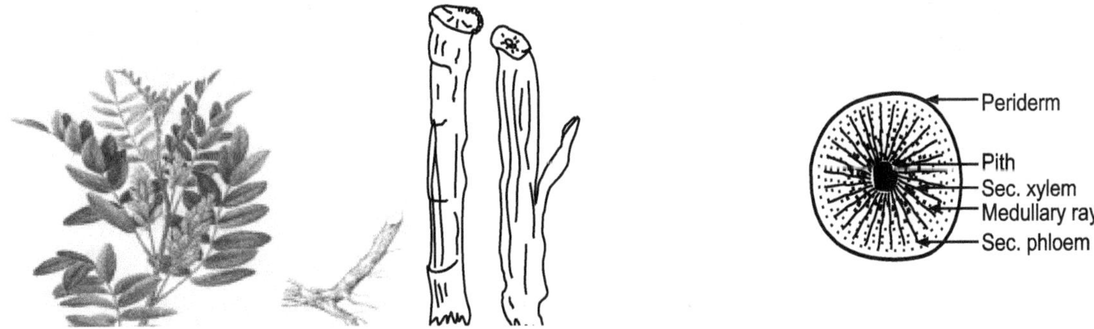

Chemical constituents: The chief consti-tuent of liquorice is a triterpenoid saponin known as glycyrrhizin (giycyr-rhizic acid), which on hydrolysis yields glycyrrhetinic acid (glycyrrhetic acid). Another important chemical consttuents are flavonoid named as liquiritin and isoliqueritin.

Uses: Licorice has expectorant, demulcent, anti arthritis, anti-inflammatory effects due to triterpenoid moiety. Italso exhibits antispasmodic and antigastric properties due to flavonoid glycoside. Flavonoid - liquiritin and isoliqueritin are antioxidant, anti-ulcer and found to be useful in skin cosmetics as depigmentation agent.

Dioscorea

Synonyms: Yam, Rheumatism root.

Biological Source: It consists of dried tubers of the plants *Dioscoreadeitoidea, D. Compositae* of family Dioscoreaceae.

GeographicalSource: India, U.S.A. and Mexico.

Morphology: Colour: Slightly brown, Odour: Odourless, Taste: Bitter, Size: Varies depending upon age of rhizomes

Microscopy: Epidermis is normally absent, thin walled parenchy-matous tissue containing cork; Stele consists of several close collateral fibro-vascular bundles, pericyclic layer

Chemical constituents: Dioscorea rhizomes contain 75% of starch. They are non-edible, since they are very bitter in taste. The chief constituents are steroidal sapogenin like diosgenin (hydrolytic product is aglycone dioscin) smilagenin, epismilagenin and 6-isomer yammogenin.

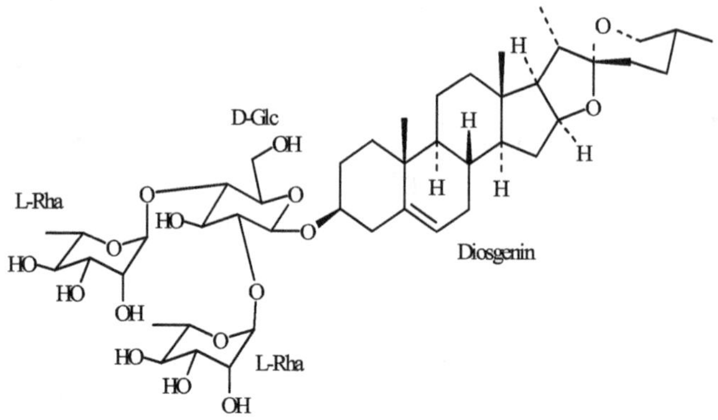

Uses: The steroidal saponins of dioscorea are used as precursor for synthesis of several corticosteroids, steroidal-hormones and oral-contraceptives. It is also useful as anti arthritic.

Cardiac Glycosides

Digitalis

Synonym: foxglove, common foxglove, purple foxglove or lady's glove, folia digitalis

Biological source: It consists of dried leaves of *Digitalis purpurea* belonging to family Plantaginaceae (formerly treated in the family Scrophulariaceae). It should not contain less than 0.3 % of total cardenolides calculated as digitoxin.

Geographical source: Native and widespread throughout most of temperate Europe. It is also naturalised in parts of North America and some other temperate regions.

Morphology

Digitalis purpurea	*Digitalis lanata*
Color: deep green	Color: deep green
Odor: tea like	Odor: tea like
Taste: bitter	Taste: bitter
Shape: ovate and lanceolate lamina with dentate margin and rounded apex	Shape: oblong and lanceolate lamina with dentate margin and rounded apex
Size: 10-20 cm long and 4-10 cm wide	Size: 22 cm long and 6 cm wide
Extra feature: Leaves are crumpled and broken	Extra feature: Leaves are crumpledand broken

Contd...

Microscopy:	Microscopy:
It is dorsiventral leaf. A transverse section shows upper epidermis and lower epidermis with anomocytic stomata and covering as well as glandular trichomes, palisade cells, palisade cells and mesophyll. Glandular trichomes are short, unicellular stalk and unicellular or bicellular head. Covering trichomes are uniseriate with colapsed cells. Digitalis is free from Ca oxalate crystals and sclerenchyma. Starch grains are present. Collenchyma is present.	It is dorsiventral leaf. A transverse section shows upper epidermis and lower epidermis with anomocytic stomata and covering as well as glandular trichomes, palisade cells, palisade cells and mesophyll. Glandular trichomes are short, unicellular stalk and unicellular or bicellular head. Covering trichomes are uniseriate with colapsed cells. Digitalis is free from Ca oxalate crystals and sclerenchyma. Starch grains are present. Collenchyma is present.

Morphology and Microscopyof Digitalis Leaf

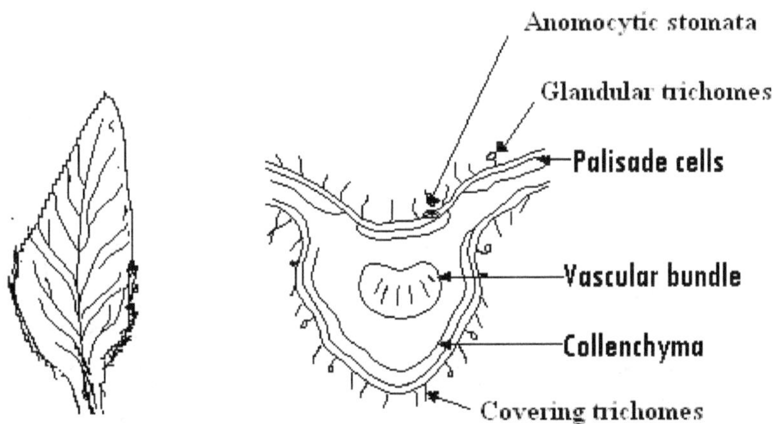

Chemical constituents:

Digitalis purpurea

1. *Primary Glycoside*
   - Purpurea glycoside A : digitoxigenin + glucose + 3 digitoxose sugar
   - Purpurea glycoside B: gitoxigenin + glucose + 3 digitoxose sugar
2. *Secondary glycosid e obtained due to hydrolysis of primary glycosides:* Digitoxin, gitoxin
3. Saponin glycosides: Digitonin, gitonin
4. Others: Verodoxin, glucoverodoxin, odoroside H,

Digitalis lanata

Primary Glycoside

➢ Lanatoside A : digitoxigenin + 2 digitoxose sugar + acetyl digitoxose + glucose

➢ Lanatoside B: gitoxigenin + 2 digitoxose sugar+ acetyl digitoxose + glucose

➢ Lanatoside C: digoxigenin + 2 digitoxose sugar+ acetyl digitoxose + glucose

➢ Lanatoside D: diginatigenin + 2 digitoxose sugar+ acetyl digitoxose + glucose

Primary glycosides are less stable and less absorbed than secondary glycosides so the concentration of secondary glycosides is more significant.

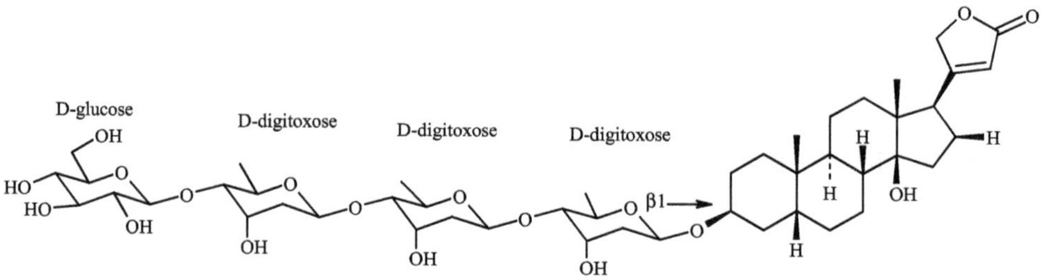

Purpurea Glycoside A = Digitoxin + D.Glucose : Digitoxin = Digitoxigenein + 3 Digitoxose sugars

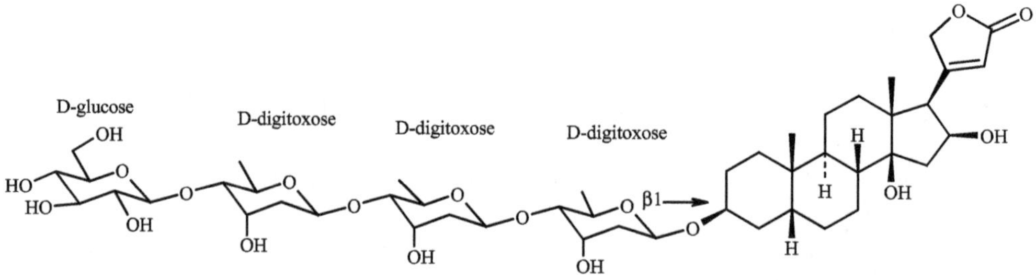

Purpurea Glycoside B = Gitoxin + D-Glucose: Gitoxin = Gitoxigenin + 3 Digitoxose sugars

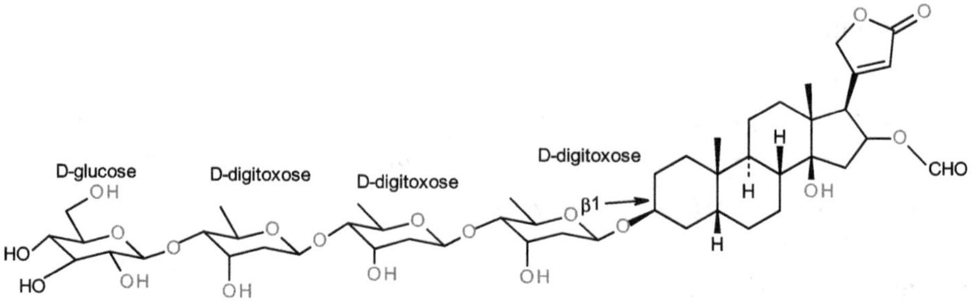

Glucogitatoxin-Gitalexin + D-Glucose : Gitaloxin = Gitaloxigenein + 3 Digitoxose sugars

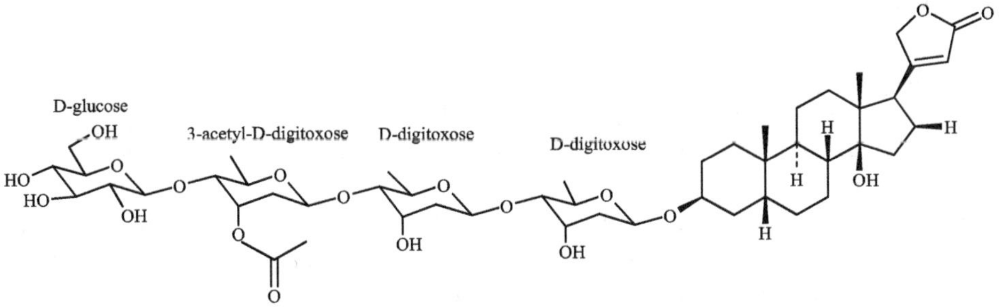

Lanatoside A = Acetyldigitoxin + D-Glucose : Acetyldigitoxin = Digitoxigenin + 2 Digitoxose sugar + 1 Acetyl digitoxose

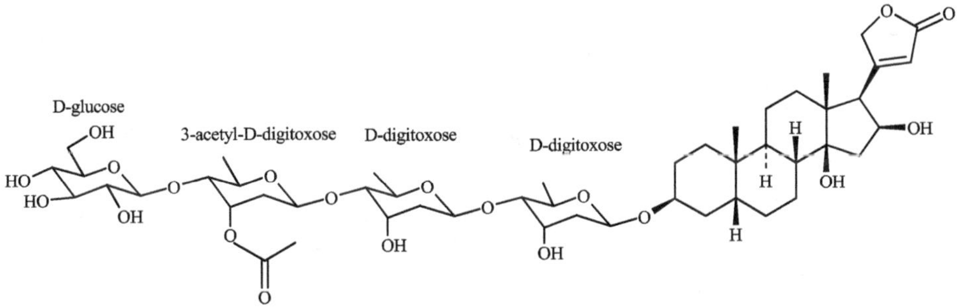

Lanatoside B = Acetylgitoxin + D Glucose :Acetylgitoxin = Gitoxigenin + 2 Digitoxose sugar + 1 Acetyl digitoxose

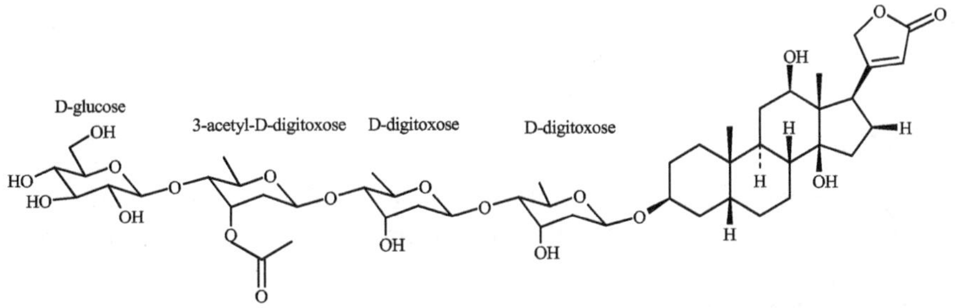

Lanatoside C = Acetyldigoxin + D-Glucose : Acetyldigoxin = Digoxigenin + 2 Digitoxose sugar + 1 Acetyl digitoxose

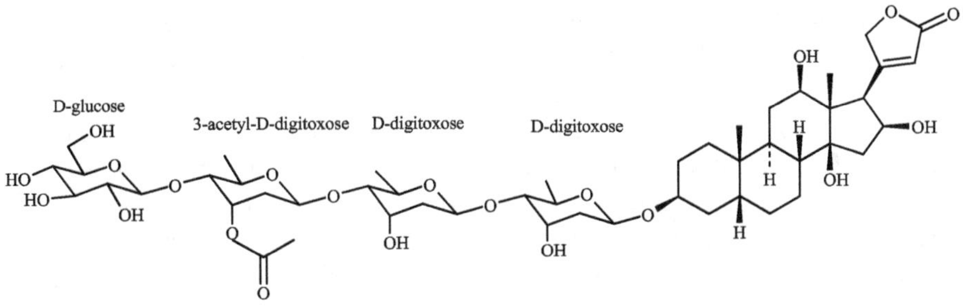

Lanatoside D = Acetyldiginatin + D-Glucose: Acetyldiginatin – Diginatigenin + 2 Digitoxose sugar + 1 Acetyl digitoxose

Chemical Tests

Kedde test	Mix 1 ml of test solution with 2 ml reagent	Bluish to purple colour
Baljet reagent	Mix 2–3 mg of sample in 2 ml sodium picrate solution.	Yellow, orange to deep red colour.
Keller-killani test for digitoxose sugar :	To the alcoholic extract of sample add 5 ml of water and 0.5 ml of strong solution of lead acetate. Filter and treat the clear filtrate with equal volume of chloroform and evaporated to yield the dry residue. Add glacial acetic acid, 0.5 ml of ferric chloride solution and 2 ml of concentrate sulfuric acid.	Initially the red -brown layer changes to blue green.
Legal test (Cardenolides)	Mix 1 ml of test solution with 2 ml pyridine + Sodium nitropruside.	Pink or red colours.
Raymond Test	Mix alcoholic extract of sample in 0.1 ml of Raymond's reagent and add 2-3 drops of NaOH solution (20%)	Violet colour changes to blue

Uses: Digitalis is very preferred drug in heart failure where it increases rate and force of contraction of heart to release the pressure of accumulated blood. It is also very useful in congestive heart failure, atrial flutter and fibrillation. Digitalis is also used as diuretic and in treatment of internal haemorrhage. The classical mechanism of action of Digitalis involves its binding to and inhibition of the plasma membrane Na+/K+-ATPase (sodium pump) to act as a positive inotropic agent. The increased intracellular sodium concentration and the increased serum potassium concentration producesnegative chronotropic and positive inotropic effect. Digitalis lanata has less cumulative effects and more potent than Digitalis purpurea.

2.4 Volatile Oils

Odorous volatile principles of plant and animal origin are called as essential or volatile oils, and the plants containing essential oils are called as aromatic plants, which are a great heritage of India. In plants, these are commonly found in families like *Labiateae, Rutaceae, Piperaceae, Zinziberaceae, Umbelliferae, Myrtaceae,* and*Lauraceae.* They are found in very few animals like musk, civet, and sperm whale. Due to their fragrance and therapeutic properties, essential oil is a multibillion dollar industry which has potential for enormous growth.

Properties: Terpenoids are complex mixtures of hydrocarbons and oxygenated compounds derived from isoprene unit (C_5H_8). But only the monoterpene (C_{10}) and sesquiterpene (C_{15})are volatile in nature. These oils are classified according to the presence of different functional groups like alcohol, aldehyde, ester, ketone, and phenol volatile oils. Some of them also contain nitrogen or sulphur. They are always unsaturated compounds having one or more double bonds. The structure can be an open chain or cyclic, with one or more carbon atom in the ring. Sesquiterpenes are linear, branched, acyclic, monocyclic, bicyclic or tricyclic unsaturated compounds. The classic examples of monoterpene essential oil composition are limonene, α-pinene, β-pinene, linalool, geraniol, citral, terpinen-4-ol, and α-terpineol. The classic examples of sesquiterpene essential oil composition are zingiberene, farnesene, farnesol, bisabolene, cadinene, humulene, and santalol.

ExtractionIsolation and Purification Methods

Generally, volatile oils are extracted from plant material by four methods, i.e., distillation, steam distillation, expression, and enfleurage, which are already discussed in chapter 1. Most oils are distilled in a single process within 2-5 hours, while a few of them require a second step to purify them through fractional distillation. Maceration, where the plant material is macerated in a warm water to release enzyme-bound essential oils, is a very specific method applicable to a limited number of essential oils like onion, garlic, wintergreen, bitter almond, etc.

Solvent extraction is one of the most promising methods of essential oil extraction, but recently, newer techniques like microwave assisted or supercritical fluid (SCF) extraction have become more popular. Microwave extraction is a mild and controllable processing tool which allows oil extraction with or without solvent or even dry extraction within a very short time. The use of SCF in essential oil extraction has gained much more attention recently due to the high selectivity in separation, no toxic residue, and thermal degradation which produces highest quality essential oil than conventional distillation methods.

There are many factors which are responsible for quality of distilled essential oils like time, temperature, pressure, and distillation equipment. Essential oils are composed of fairly delicate components which can be altered, destroyed or lost in their activity by improper methods of extractions.

Dry Distillation Method

Dry distillation, which also called as destructive distillation method, heats the plant material without water, and the vapours of oil condense to give total volatile oil. The drawback of this method is smell of roasting in the finished product due to direct heating of raw material.

Solvent Extraction Method

Many of the chemical components of plant parts like flowers are too delicate and easily denatured by the high heat and temperature used in steam distillation. Hence, for thermolabile components, extraction by using solvents such as petroleum ether or hexane is a very good alternative. The major problem with this method is the final extract of ether or hexane, known as "concrete", contains essential oils along with waxes, resins, and other lipophilic components, and thus, it requires one further step of purification. Hence, ethanol is used to extract the fragrant oils from the concrete which is then removed by a second distillation, leaving behind the purified essential oil which is called as absolute.

Rectification

This is the most general purification method where fractionating columns are used to distill mixtures of volatile components so as to separate the fractions, based on their boiling points. This is applicable to small scale laboratory as well as large-scale industrial distillations.

General Method of Extraction of Volatile Oils

Select the appropriate Clevenger apparatus for volatile oils having density either higher than or lower than water. Take the accurately weighed quantity of fresh or dried plant material consisting of the flowers, leaves, wood, bark, roots, seeds, or peel (e.g. eucalyptus leaves, lemongrass leaves, clove buds etc). Reduce the material to coarse sized particles. Place the material in RBF of Clevenger apparatus. Add sufficient quantity of water. Attach the assembly as in the figure. Heat the mixture for 1 hr. Vapours of both water and volatile oil will pass through the condenser and then the collector tube. Allow the mixture to cool and separate. Read the volume of oil directly from the graduated tube.

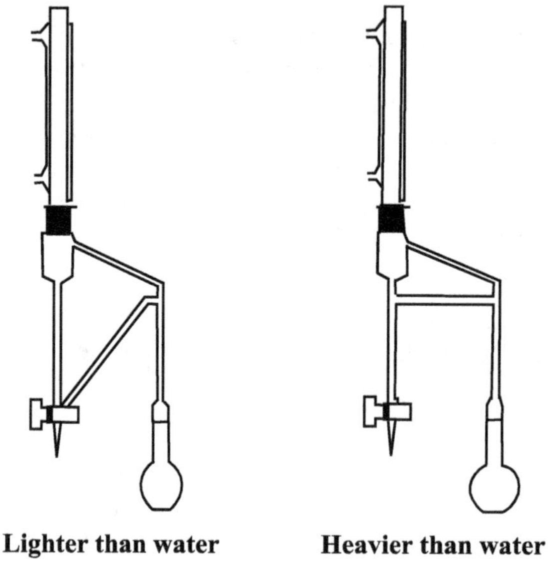

Fig.2.5 Clevenger apparatusfor volatile oil determination

Mentha

Synonym: Peppermint, Peppermint oil, Oleum menthapiperata, colpermin, mentha oil,

Biological source: The oil is obtained by steam distillation from leaves and fresh flowering tops of the plant *Mentha piperita*, family –labiatae.

Other species: M.aquatica (watermint); M.spicata(spearmint); Varieties: American mentha, Black mint and white mint.

Geographical source: Europe, japan, England, USA, India

Cultivation and Collection

Propogation: Vegetative by suckers

Climate: Temp. Between 15-25 c,

Altitude: 250-400m

Soil: Sandy loamy, Neutral pH

Fertilizer: Superphosphate, potash

Pesticide: Mercury compound, thiodan, BHC.

Harvesting: At flowering stage: dried in shade or artificially

Preparation of oil: Steam distillation followed by rectification.oil is insoluble in water, Lighter than water, Percentage of oil: 1-3%

Morphology: Peppermint leaves: serrate margin, Acute apex, Dark green with purple tinge; Steam: smooth decussate, Creeping rhizome; Distilled oil: Pale yellow-greenish colour, Odour: Pleasant, taste: Pungent.

Microscopy: Upper and lower epidermis: Leaf; Diacytic stomata, Palisade, spongy parenchyma covering as well as glandular trichomes; No ca-oxalate crystals presents; Volatile oil is secreted from beneath cuticle of multicellular glandular trichomes.

Chemical constituents: It contains **1-3%** Volatile oil constituting limonene (1.0-5.0%), cineole (3.5-14.0%), menthone (14.0-32.0%), menthofuran (1.0-9.0%), isomen-thone (1.5-10.0%), menthyl acetate (2.8-10.0%), isopulegol (max. 0.2%), menthol (30.0-55.0%), pulegone (max. 4.0%) and carvone (max. 1.0%).It also contains Flavonoids (Liteolin, hesperidin, rutin, menthoside, asorhoifolin), Phenolic acid (Chlorogenic acid, rosmarinic acids), Triterpenoids (alpha-amyrin, ursolic acid, caretenoids, choline and betaine) and Sesquiterpene alcohol.

Piperitenone   Piperitone   Menthone   Menthol

Pulegone   Menthofuran   Menthyl acetate

Uses: It is used as carminative, flavouring and aromatic agent, analgesic, anesthetic, antiseptic, astringent, carminative, decon-gestant, expectorant, nervine stimulant, stomachic, inflammatory diseases, ulcer and stomach problems.

Allied drugs: Japanese peppermint-*Mentha arvensis(*85% Menthol)

Cinnamon

Synonym: Cinnamon bark, Ceylon cinnamon, Kalmi-Dal-chini, Common cinnamon

Biological source: Oil is obtained from inner bark of plant Cinnamomum zeylanicum of Family Lauraceae. Genus Cinnamomum contain 250 species

Geographical source: native of India, Sri Lanka, Malaya

Morphology: Leaves: shiny green above and white underside spicy odour; Flower: Yellow white; Bark: Outer yellowish brow-inner dark yellowish brown; Taste-Aromatic and sweet with warm sensation; odour: fragrant; fracture-splintery, wavy longitudinal striations, compound quill: shape

Microscopy: Due to inner bark, cork and cortex absent, presence of pericyclic fibre, sclerides, medullary rays, phloem fibers, secretary cavity, starch grain, Acicular Ca-oxalate crystals also contain mucilage.

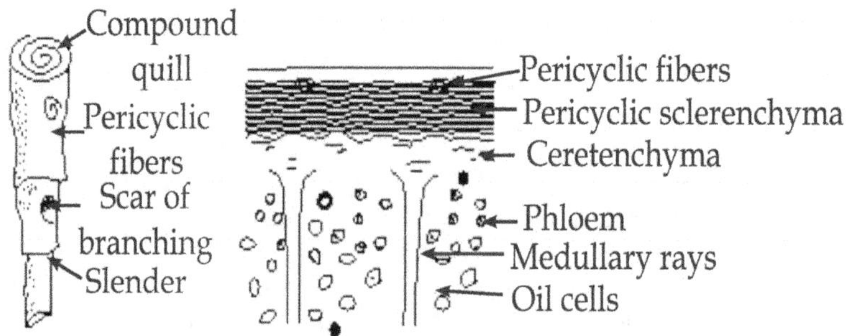

Chemical constituents: Bark contains 0.5-1.0%volatile oil composing 60-70%, cinnamaldehyde, 5-10% Eugenolandother constituents like benzaldehyde, cuminal-dehyde, Phellandrene, Pinene and Cymene.

eugenol

Uses: It is used as Flavouring agent in baked foods, candies, soft drinks, table sauce, confectionaries.It is Carminative, Stomachic, Mild astringent and powerful bactericide. It is preferred ingredient indentrifries and perfumes.

Adulterant: Jungle cinnamon: dark in colour, bitter in taste; Java cinnamon: Presence of tubular ca-oxalate crystals, double quill form; Saigon cinnamon: bark is grayish brown

Sesquiterpenoides

Clove

Synonyms: Caryophyllum; Clove flower; Clove buds

Biological Source: Clove consists of dried flower buds of *Eugenia caryophyllus* of family Myrtaceae.It is one of the highest volatile oil (15-20%) containing crude drugs.

Geographical Source: It is indigenous to Amboyna and Molucca islands and cultivated chiefly in West Indies, Sri Lanka and India (Nilgiri, Tenkasi-hills and in Kanyakumari, Tamil Nadu state, Kerala).

Morphology: Colour: Crimson to dark brown; Odour: Slightly aromatic; Taste: Pungent and aromatic followed by numbness and Size: About 10 to 17.5 mm in length, 4 mm in width, and 2 mm thick; Shape: Hypanthium is surmounted with 4 thick acute divergent sepals surrounded by dome shaped corolla. The volatile oil of the drug contains Oil of clove is colourless to pale yellow in colour. It becomes thick and darker in colour on storage.

Microscopy: Transverse section through hypanthium shows presence of epidermis with thick cuticle, straight walled cells, large anomocytic stomata, ovoid and schizoly-sigenous oil gland, Phloem fibres, Cluster crystals of calcium oxalate and stone cells. Starch grains are absent.

Starch, calcium crystals and stone cells are absent in transverse section through ovary region.

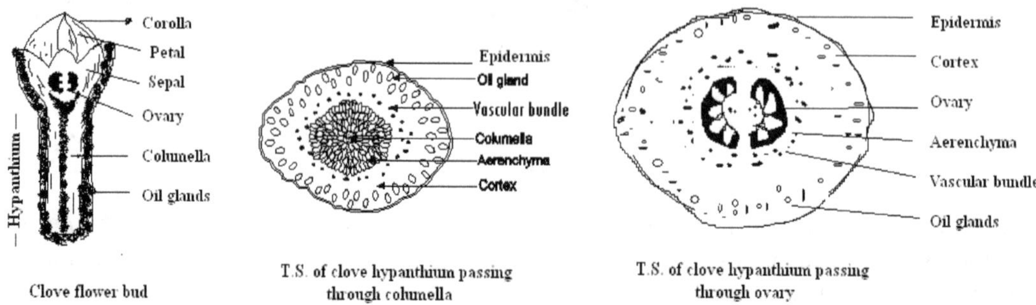

Chemical constituents: It contains about 15 to 20% of volatile oil composing eugenol (about 70 to 90%), eugenol acetate, caryo-phyllenes and small quantities of esters, ketones and alcohols. It also contains 10% to 13% of tannin (gallotannic acid), resin, chromone and eugenin.

Eugenol

Eugenol acetate

Caryophyllene

Uses: It is used as a aromatic, flavouring agent, carminative, dental analgesic, stimulant and 'antiseptic. The oil is used in the manufacture of vanillin.

Adulterants: Mother cloves and Blown cloves with less oil content, Clove stalks, Exhausted cloves

Fennel

Synonym: Fennel fruits, *Fructus foeniculum*, Common fennel.

Biological source: Oil is obtained from dried ripe fruits of plant known as *Foenoculum Vulgare*, Family-Umbelliferae. It should not contain less than 1.4% of volatile oil.Dulce variety gives sweet oil.

Geographical source: Indigenous to southern Europe, cultivated in Russia, Germany, France, Japan, India (Gujarat)

Morphology: Fruit: green to yellowish; Odour-sweet aromatic; Taste-Bitter and Aromatic; Flower-Yellow in colour, seed greenish in colour, fruit in the form of cremocarp

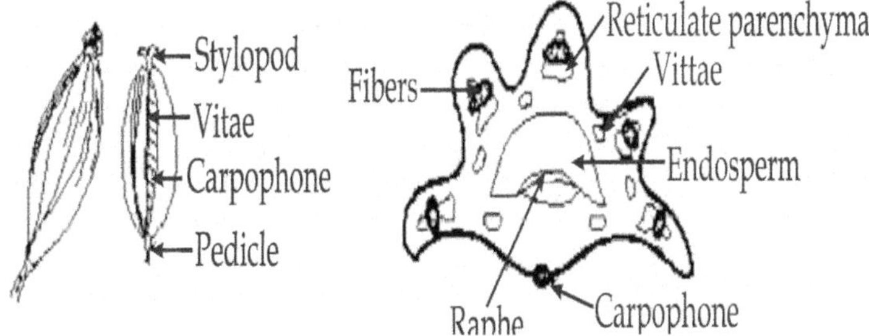

Microscopy: Presence of anomocytic stomata on epidermis of pericarp and mesocarp containing lignified reticulate parenchyma, rosette-ca-oxalate crystals present. Parquetry cells, vittae, aleurone grain are present but starch grainare absent

Chemical constituents: It contains 2.5-5% of Volatile oil constituting pungent principle Fenchone and sweet principle Anethole along with alpha-pinene, camphene, d-alpha-phellandrene and dipentene. Dulce variety contains high amount of anethole and absence of fenchone so it is of finest quality and odour and flavour is very sweet

Anethole   (+)-Fenchone   Camphene

Use: It is used asCarminative, condiment, Aromatic, Stimulant, Expectorant and flavouring agent in food, liquors.

Coriander

Synonym: Coriander fruits

Biological source: Oil is obtained by continuous distillation of dried ripe fruits of *Coriandrum sativum* of Family umbeli-ferrae.

Geographical source: Indigenous to mediterraneous countries, cultivated inEurope,Asia, India-Andhra Pradesh, Maharashtra, Jammu and Kasmir

Morphology: Coriander fruits: Straw yellow to brownish yellow entire cremocarp can be converted to two hemispherical mericarp, presence of 10 primary ridges and 8 secondary ridges; Odour: Aromatic, Taste-Spicy and characteri-stic

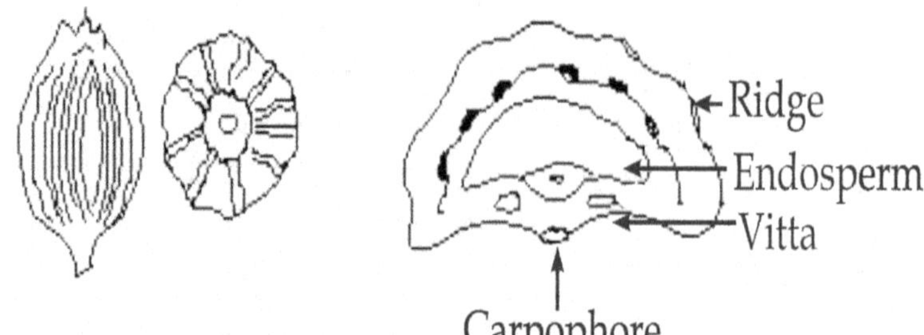

Microscopy: T.S shows two equal halves of mericarp consisting epicarp, mesocarp, stomata, trichomes, lignified reticulate parenchyma, Aleurone grains and vittae. Starch grains are absent

Chemical constituents: it contains **0.4**-1.8% of volatile oil composing 65-70%, Linalol, Coriandrol-alpha-pinene, Limonene, r-terpinene, p-cymene. It also contains Flavonoids, Coumarins, Asocoumarins and Fixed oil

Uses: It is used as aromatic, Carminative, stimulant and flavouring agent. In the form of Infusion oil is used to prevent flatulence, stomach cramps, gripping. It is aphrodisiac, Stimulant, appetizer and useful in diarrhea, Dysentry, colitis, rheumatic diseases. Linalool is used as raw material for the production of citral and beta- ionone.

Substitute: Bombay coriander

2.5   Tannins

The term tannin (from *tanna*, German word for 'oak') refers to the use of wood tannins from oak in tanning animal hides into leather. However, the term "tannin" by extension is widely applied to any large polyphenolic compound containing sufficient hydroxylsand other suitable groups (such as carboxyls) to form strong complexes with proteins and other macromolecules. Tannins molecular weights ranges from 500 to over 3,000 (gallic acidesters) and up to 20,000 (proantho-cyanidins).   Tannins are classified as ergastic substances, i.e., non-protoplasm materials found in cells. **Tannin** is an astringent, bitter, plant polyphenolic compound that binds to and precipitatesproteins and various other organic compounds including amino acids and alkaloids.

Classification of Tannins

Class	Hydrolysable tannins (divided into simple gallotannins and complex ellagitannins)	Non-hydrolysable or condensed tannins oq2r flavolans or proanthocyanidins	Phlorotannins
Basic nucleus	Gallic acid	Flavn-3,4-Diols	Phloroglucinol
Chemistry	Glucose linked gallic acid or ellagic acid in the form of galloylglucose depside or hexahydroxydiphenic acid	High molecular weight oligomers of catechins and flavan-3,4-diols	These are polymers of Phloroglucinol and restricted to be present in brown algaes
Sources	Dicot Plants	Plants	Algae

Extraction

Extract the powder crude drug n-hexane or petroleum ether to remove impurities. Then extract marc with acetone. Add diethyl ether and separate acetone layer to get tannins. Extraction of tannins requires skills to separate other polyphenolic compounds too.

Catechu

Common name: Kattha, Cutch, Khadir-catechu, Catechu.

Biological source: It consist of dried aqueous extract of plant *Acacia catechu* Wild *Acacia chundra*wild family Leguminoseae.

Geographical source: India and Myanmar

Morphology: Color: Light brown to black, odour: none, taste: astringent, size: About 2.2 to 5 cm, shape: cube or irregular fragments of broken cubes or brick shaped pieces, extra features: cubes as well brick shaped pieces of catechu shows the presence of vegetable debris, breaks with short fracture. The broken fracture pieces are angular with pale cinnamon brown in colour.

Manufacturing Kathha/Khair/Catechu

Kattha and cutch are extracted from wood of Khair (*Acacia catechu*) tree. Another source for Kattha substitute is *Uncaria gambier which* is found in Malaysian region and imported in India.Kattha is widely consumed all over India as an applicant in paan (Betel leaf). It also used as astringent. Cutch is a bye product of kattha manufacturing which is used in tanning industry and as an additive and preservative in many industries.

Extraction: Cut the heart wood of acacia tree into fine chips. Boil chips with water for about three hours. Extract three times. Filter from muslin cloth and concentrate in an open pan on fire and then keep in shade to facilitate crystallization of kattha for about two days. After complete crystallization, pass the curd like mass through frame and plate-type filter press operated manually. Then wash with cold water to improve the quality of kattha. Now place on wooden frames with canvas cloth to separate traces of cutch. Finallykattha is cut into uniform tablets with the help of wire cutter or knife and dry in sheds. Concentrate mother liquor after removal of kattha in an open pan till it becomes viscous and then pour in wooden frames for drying. This dried material is cutch. Yield: kattha: 5% and cutch: 14%

Chemical Tests:

➤ Black catechu treated with vanillin hydrochloric acid produces pink or red colour.

➤ Match stick test: Tannin extract on stick dipped in HCL and heated near flam produces purple magenta colour.

➤ Aqueous extract of catechu treated with lime water gives brown colour

➤ Dilute solution of drug treated with Ammonium sulphate gives green colour.

Chemical constituents: It contains catechin, catechu tannin acid, flavonid- quercetin, and gum.

Uses: It is used externally as astringent to boils, eruptions, and ulcers. It is also used in cough and diarrhea. Both catechus are ingredient of many breath freshening mixtures, pan-masala and Gutka products.

Pterocarpus

Synonym: IndianKino, Bijasal, Vijayasagar, Bibla, Malbar kino

Biological source: Dried bark of Pterocarpus marsupium plant belonging to family Fabaceae

Geographical source: India, Srilanka and Nepal.

Chemical constituents: Earlier researchers recognized the plant P. marsupium as a very rich source of flavonoids and polyphenolic compounds. All the active phytoconstituents of P. marsupium were thermostable. It contains pterostilbene (45%), alkaloids (0.4%), tannins (5%) and protein.19Theprimary phytoconstituents were liquiritigenin, isoliquiritigenin, pterostilbene, pterosupin, epicatechin, catechin, kinotannic acid, kinoin, kino red, β-eudesmol, carsupin, marsupial, marsupinol, pentosan, p-hydroxybenzaldehyde

Uses: *P. marsupium* has been traditionally used in the treatment of leucoderma, elephantiasis, diarrhea, cough, discoloration of hair and rectalgia. It is nontoxic and useful in jaundice, fever, wounds, diabetes, stomachache and ulcer. Its heartwood possesses anti-inflammatory, astringent, anodyne and antidiabetic properties and also cataract, hypertriglyceridemia, cardiotonic, hepatoprotective activity and as a selective inhibitor of COX-2.

2.6 Resins

Resins are highly complex entities compromising of a complex mixture of either acid, phenol (resinotanol), alcohol (resinols), and ester or inert neutral compounds (resene). These are water soluble amorphous moieties. They are rich in C, H, and O but devoid of N. These are secreted by specialised cells or ducts, known as schizogenous ducts. A few of the resins like cannabinoids are secreted from the trichomes, i.e., plant hairs. Resins are generally classified as follows on the basis of composition:

➤ *Oleo resins*: Oleoresins are mixtures of oil and a resin. e.g turmeric, ginger, copaiba

➤ *Gum resins:* Gumresins are mixtures of gum or mucilaginous substances and a resin. e.g ammoniacum

➤ *Oleo*-gum resins: *Oleo*-gum resins are mixture of oil, gum and resin. e.g. myrrh, asafoetida,

➤ *Balsams*: Balsams are containing benzoic acid, cinnamic acid or their esters and resin. e.g. tolu balsam, peru balsam, storax

General extraction method

Resins are insoluble in water but soluble in organic solvents, fixed oils, volatile oils, and chloral hydrate solution. The extraction of resins is a very simple process as most of the resin exudates are used in their natural form which are called natural resins, (Example - balsams and turpentine.) Some resins need purification process like solvent extraction, and evaporation, (Example - jalap, podophyllum).

Benzoin

Synonym: Sumatra benzoin, Loban

Biological source:balsamic resins obtained from *Styrax benzoin* or *Styrax paralloneurus* (Sumatra benzoin) and also contain balsamic resin from *Styrax tonkinesis*(Siam benzoin) of familystyraceae

Geographical source:South eastern Asia (Sumatra) and Thailand and Vietnam (Siam).

Morphology:

Sumatra benzoin	Siam benzoin
Colour : grayish-brown or grey	Colour : yellowish brown to rusty brown
Taste : sweetish and slightly acrid	Odour : agreeable and vanilla like
Odour : aromatic and characteristic	Taste : sweetish and slightly acrid
Shape: tears	Shape: hard and brittle masses

Chemical constituents:

- Sumatra benzoin contains free balsamic acids (benzoic and cinnamic acid),triterpenoid acids-summaresinolic acid and siaresinolic acids,styrene (2.3%), cinnamic acid (3.5%) and benzyl cinnamate (3.3%)

- Siam benzoin contains coniferyl benzoate, styrol, vanillin and phenyl propyl cinnamatebenzyl benzoate (76.1–80.1%) for the two oils and benzoic acid (12.5%), methyl benzoate (1.5%) and allyl benzoate (1.5%)

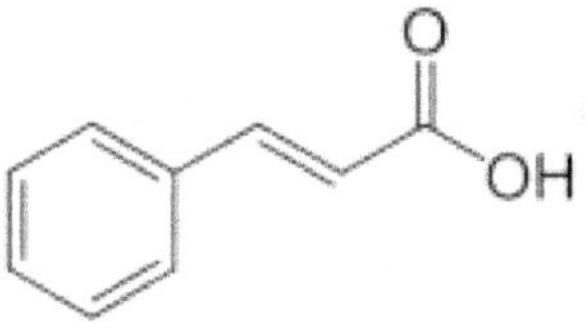

Cinnamic acid 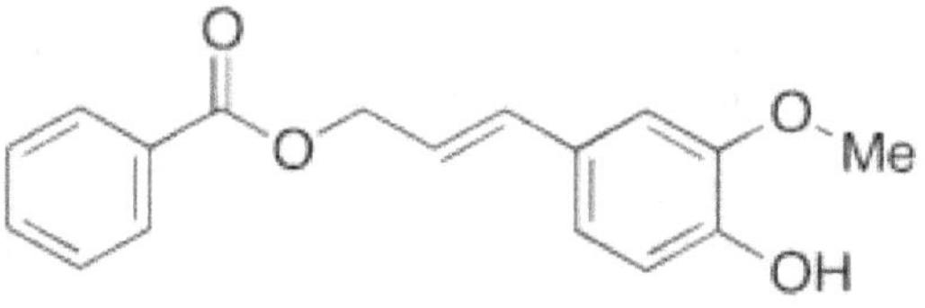   **Coniferyl benzoate**

Uses: It is topical protectant and antiseptic and hence find its uses in incense, soaps, perfumes and other cosmetics. It frequently employed as expectorant, carminative and diuretic. Internally it is used in treatment of upper respiratory tract infection.

Myrrh

Common name:-Gum myrrh, Bol, Myrrha

Biological source:- It is an Olio –gum resin obtained from *Commiphoramolmol*of family burseraceae

Geographical source:-North East Africa and Sothern Arabia

Description: color-externally it is reddish brown and internally brown, odour-aromatic and agreeable, taste –acrid, size-1.5 to 3.o cm in diameter, shape-rounded or irregular tears, extra feature-fracture surface is granular, it is brittle and shows translucent surface

Chemical constituents: Itconsists of 60% of resin composing α, β and γ commiphoric acids. It also contains about 10% volatile oil composing eugenol, cuminaldehyde and limonine. Other constituents are pyrocatechin and procatechuic acid. Myrcene and a-camphorene, as well as a few steroids including Z-guggulsterol and I, II, III guggulsterol are also present

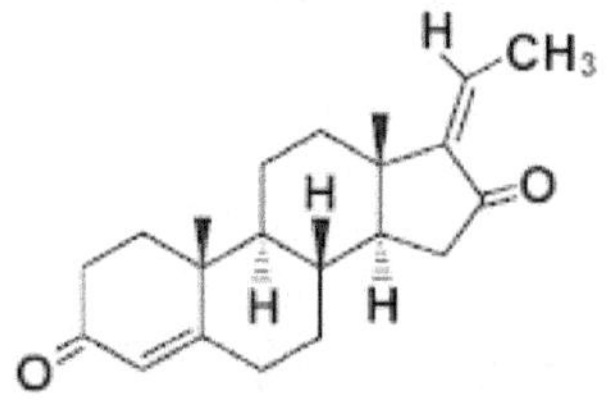

Z-guggulsterone 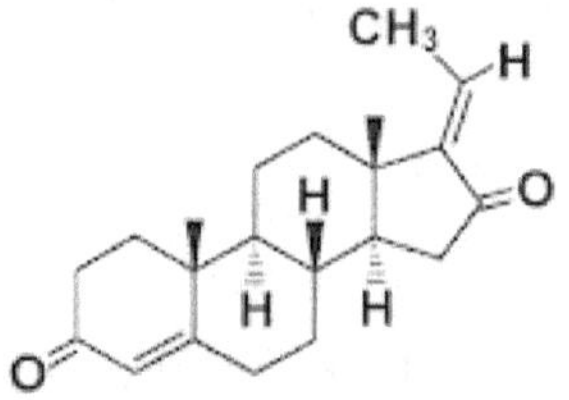   *E*-guggulsterone 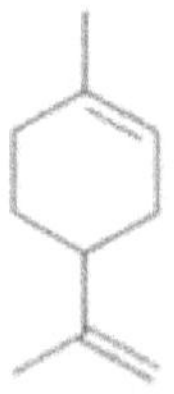   limonene 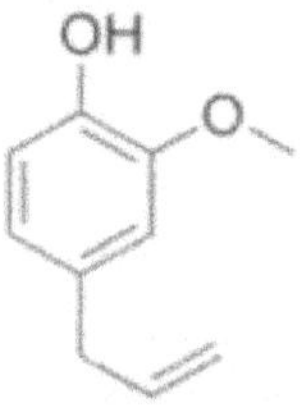   eugenol

Uses: It is frequently used as stimulant, carminative and anti septic. In Ayurveda it is one of *Rasayana* means adaptogenic drug. Due to its astringent effect it is employed in mouth washes and gargles. It has also very good application in perfumery.

Guggul

Common name: Scented bdellium, Gum Guggul

Biological source: It is oleo-gum resin obtained by making deep incision at the basal part of steam bark *Commiphoraweightii* of family Burseraceae.

Geographical source: Ethiopia, Somalia, Kenya, Zimbabwe India.

Description: Colour: Brown to pale yellow or dull green, odour: Agreeable Aromatic and balsamic, taste: bitter, size: 0.5 to 1.00 to 2.5 cm in dim, shape: Rounded or irregular masses Tears are somewhat transparent. Solubility: Partly soluble in alcohol. When triturated with water it forms white emulsion.

Chemical constituents: Resin portion of guggul contains steroid diterpenoides named as Z guggulosterone, E guggulosteron and steroids like guggulosteroid 1, 2, 3.

Uses: It is extensively used as anti inflammatory, anti rheumatic, hypo-lipidemic and hypocholestenic agent

Adulterants:*Commiphoramolmol, C.abysinnica, C.roxburggi, Boswellia sarata.*

Colophony

Common name: Rosin, Rosina, Colophonium resin, Wood rosin

Biological source: It is the residue remained after distillation of crude oleo resin of *Pinus palustris and other Pinus species* of family pinaceae.

Geographical source: North America, Northern Europe, Pakistan and India.

Description: Color: pale yellow to yellowish brown, odour : faint, Taste: Turpentine like, shape: angular, translucent masses of various sizes, extra feature: brittle and readily fusible with glossy appearance.

Chemical constituents: It consists of 90% resin composing abietic acid, dihydroabietic acid and 0.5% volatile oil (Pinene).

Abietic acid

Uses: Therapeutically it is stimulant and diuretic. But mainly used in preparation of ointments, plasters, varnishes, cements, paper, soaps, printing inks and wood polishes. Its derivatives are used in tablet film, enteric coating purpose and to formulate microcapsules and nanoparticles.

Asafoetida

Synonyms: Devil's dung, Gum asafoetida

Biological source: It is an oleo gum resin obtained from the rhizome and root of *Ferulanarthex* Bioss and other species of Ferula like *Ferulafoetida* family of Umbelliferae

Geographical source:*Ferula narthex* grows abundantly in the villages of Kashmir in Baltistan, and other species like *Ferulafoetida* grows in Persia, Kandahara, and Afganistan

Method of preparation: It occurs in three forms for marketing: paste, tear and mass. In the cortex of the stem and also in the root there are numerous large schizogenous ducts filled with whitish gum resinous emulsions. After about five years when the roots has stored sufficient reserves.

Description: Asafoetida tears are generally yellow to gray but on storage turns to dark colors. Taste is bitter and acrid. Odor is intense and persistent alliaceous.

Chemical constituents: It contains about 40–64% resin, 25% endogeneous gum, 10-17% volatile oil, and 1.5–10% ash. The resin portion is known to contain asaresinotannols 'A' and 'B', ferulic acid and umbelliferone.

Umbelliferone   Fenulic acid

Uses: is used as a digestive aid, in food as a condiment, and in pickles. It is also used traditionally as antiflatulent agent to remove gases as well as to treat constipation.

Ginger

Synonym: Adrak, Sunth, Inchi

Biologicalsource: It is rhizome of *Zingiber officinale* of family zingiberaceae.

Morphology: It is buff color with aromatic odour and pungent taste.Longitudinal striations and occasional projecting fibres are seen on surface. It has nodes and internodes, scale leaves, axillary buds, adventitous roots and a terminal bud.

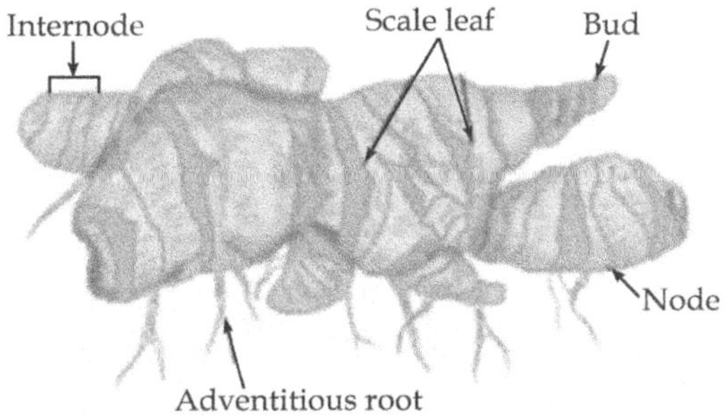

Microscopy: Transverse section shows cork of irregularly arranged cells in outer layer and radially arranged cells in the inner layer. The thin walled parenchymatous layer of cortex has abundant starch grains, reticulately thickened vessels, bright yellow ovoid to spherical cells of oleo-resin in parenchyma cells, less lignified large dentate walled fibers, closed-collateral fibro vascular bundle

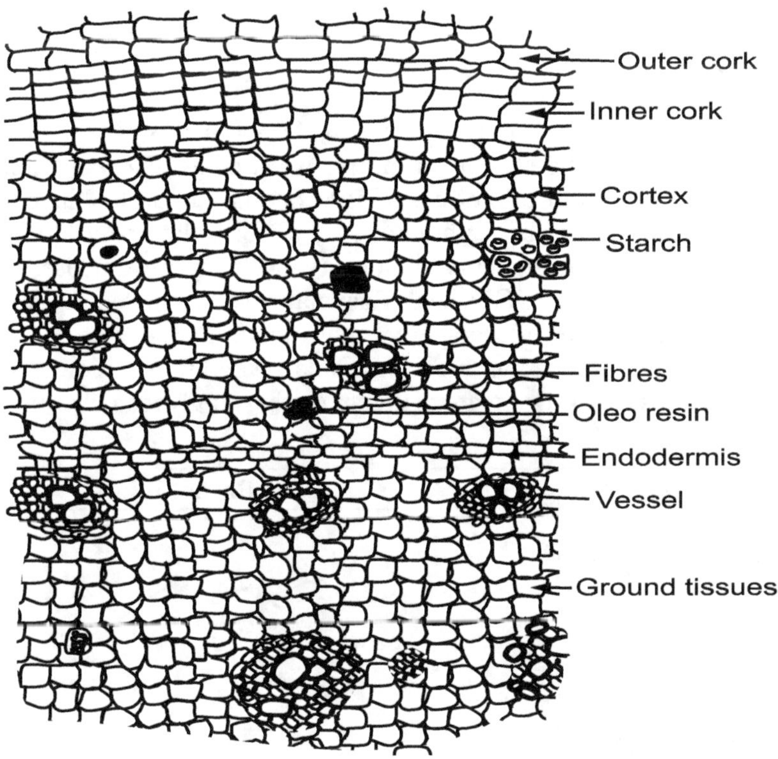

Chemical constituents: Ginger is a rich source of volatile oil. Zingiberol, zingiberene, phellandrene and linalool are important constituents of the oil. They account for the aroma of the drug. The pungency of the ginger is due to gingerols and shogoals. Investigations have shown gingerol and shogoals to be mutagenic.

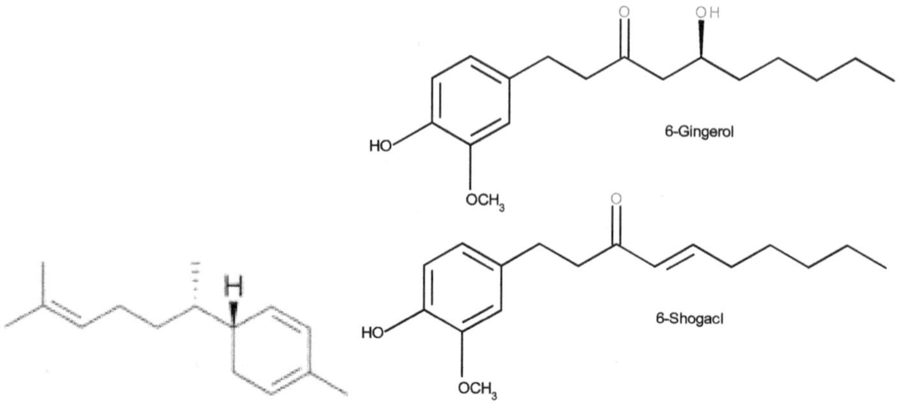

Volatile compounds: Zingiberene Non-volatile compounds

Uses: s used as carminative, expectorant, astringent, anti emetic, antiviral, hypolipidemic and in migraine. It is also reported to possess abortifacient activity.

2.7   Glycosides

Glycosides can be defined as "organic compounds which on hydrolysis give one or more sugar moieties along with non-sugar moiety. Sugar moiety is called as *glycone* and non-sugar moiety is called as *aglycone*. The therapeutic effect of aglycone moiety is facilitated by the nature of glycone moiety. Aglycones may be anthraquinone, sterols, flavonoids, terpenoids, steroids, cyanogenetic, and isothiocyanate moieties, while glycone may be glucose, galactose, xylose, mannose, digitoxose, thevatose, etc. Glycosides have always fascinated researchers due to their structural diversity and a number of therapeutic, nutritional and cosmetic applications. Due to complex chemistry, polar nature and high molecular weight, isolation and purification of glycosides is a difficult task. Following are different classes of glycosides:

Class with example	Basic Moiety
Anthraquinone ➤ Sennoside A, B, C, D from senna ➤ Palmidin A, B, C, D from rhubarb ➤ Aloin, Barbaloin from aloe ➤ Cascarosides from Cascara	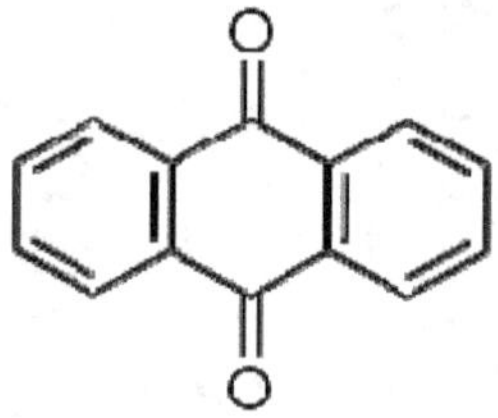

Contd...

Flavonoid ➢ Hesperidin from citrus fruits ➢ Rutin from Buckwheat	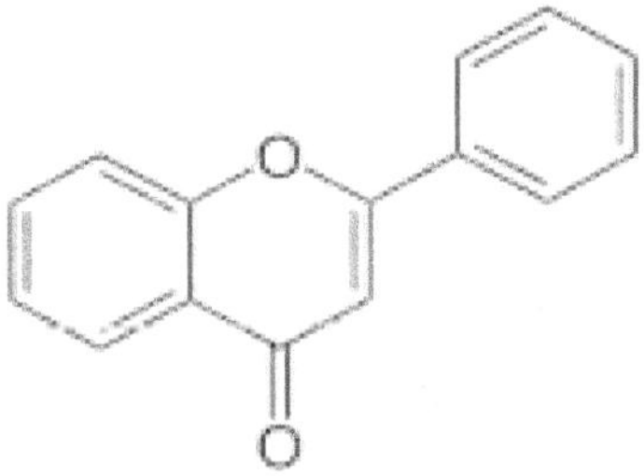
Sterol ➢ Cardenolide digitoxin, gitoxin, digoxin from digitalis ➢ Cardenolide K-stropanthin and G-stropanthin from stropanthus seeds ➢ Cardenoide thevetin-A, thevetin-B from thevetia seeds ➢ Bufadienolids scillaren A from Squill	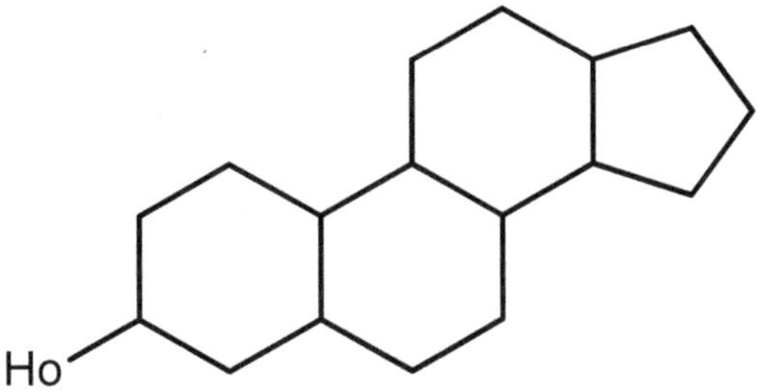 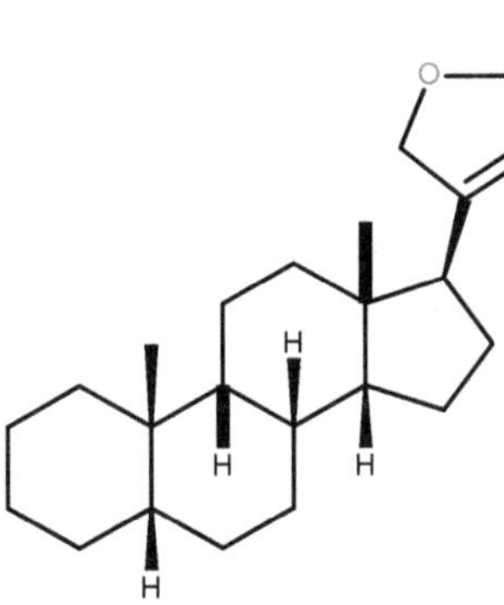 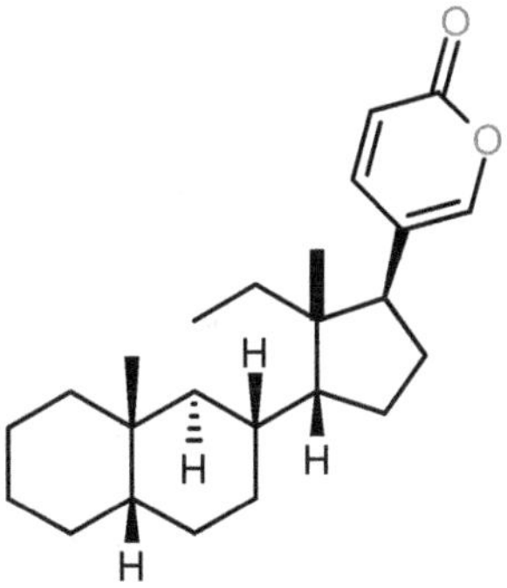 Cardenolide Bufadienolide
Saponin ➢ Steroidal saponins diosgenin from dioscorea ➢ Steroidal saponins ginsenosides from ginseng ➢ Triterpenoidal saponins glycyrrhizin from licorice ➢ Triterpenoidal saponin senegin from senega	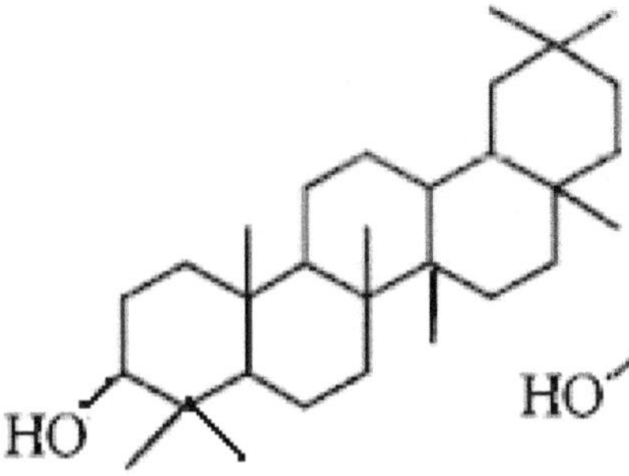 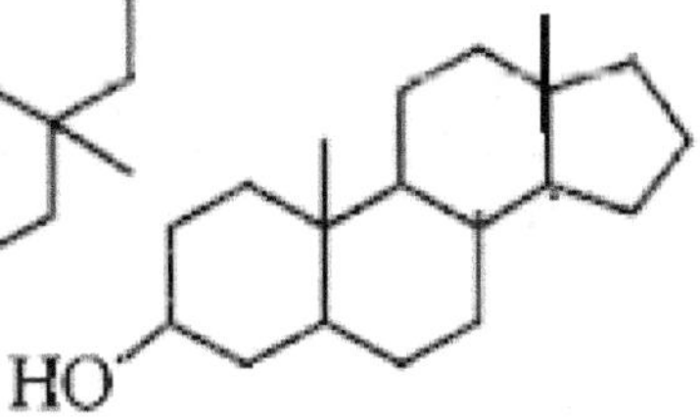 Triterpenoid saponin Steroid saponin

Contd...

<table>
<tr><td colspan="2">Types of triterpenoid saponins
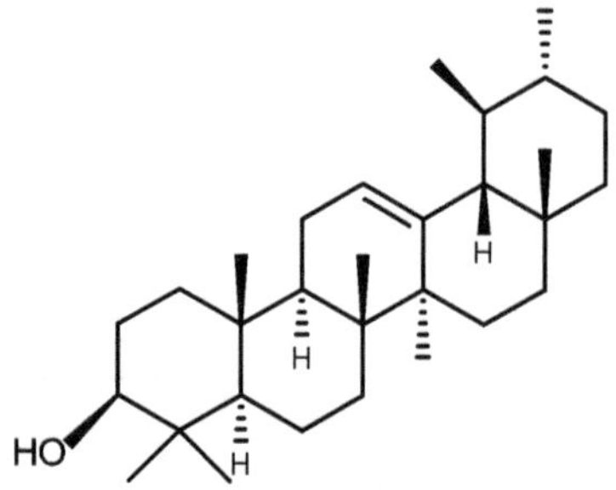
Alpha-amyrin
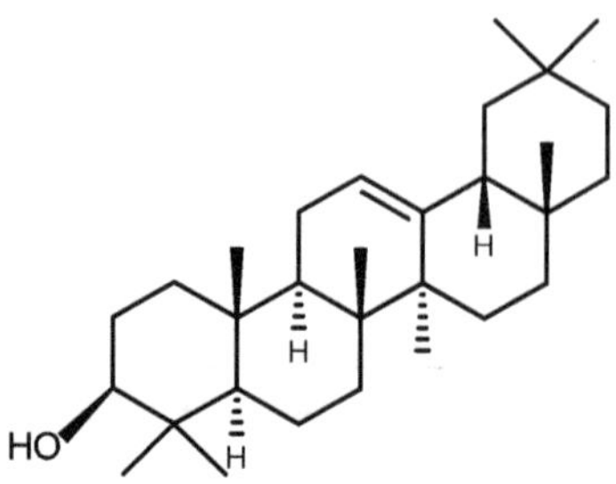
Beta-amyrin
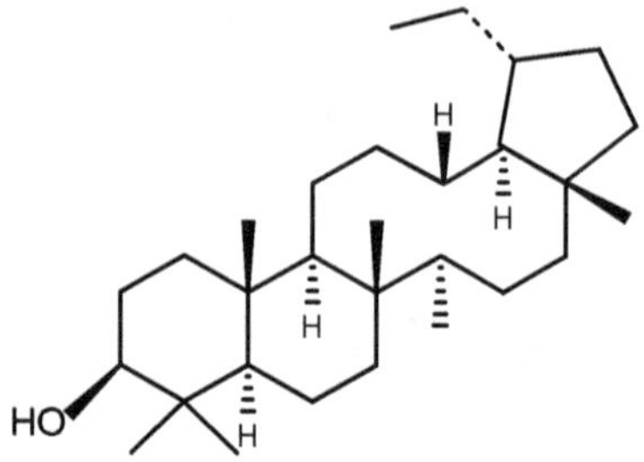
Lupeol</td></tr>
<tr><td colspan="2">Types of steroid saponins
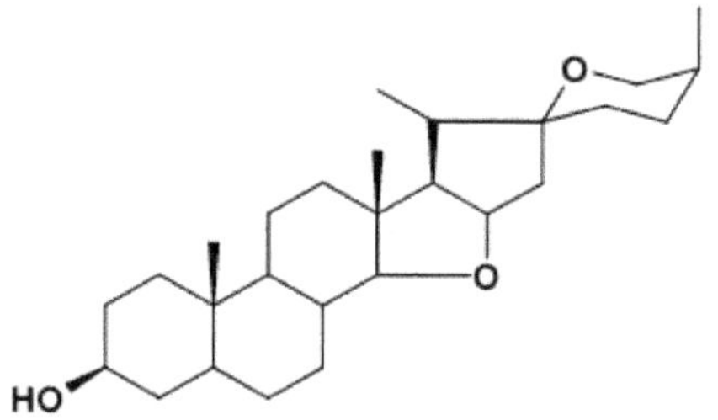
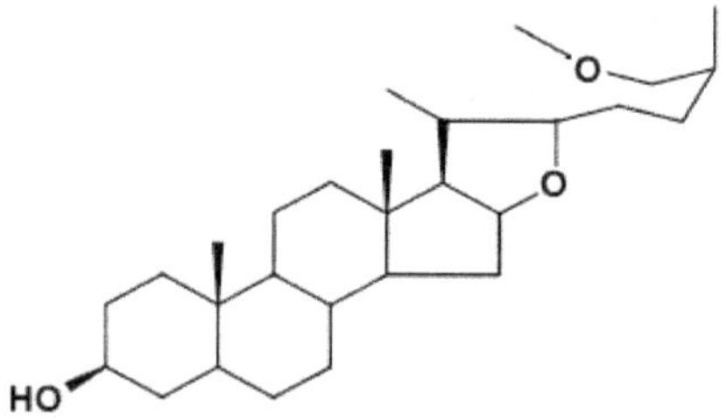
Spirostanol Furostanol</td></tr>
<tr><td>Coumarin
➢ Khellin, visnagin from Visnaga
➢ Bergapten, xanthotoxin from Ammi
➢ Psoralen from Psoralea</td><td>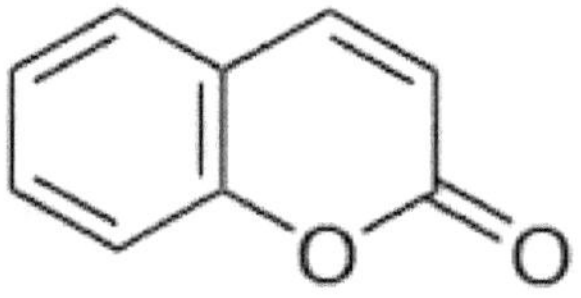</td></tr>
<tr><td>Isothiocynate
➢ Sinigrin from black mustard
➢ Sinablin from white mustard</td><td>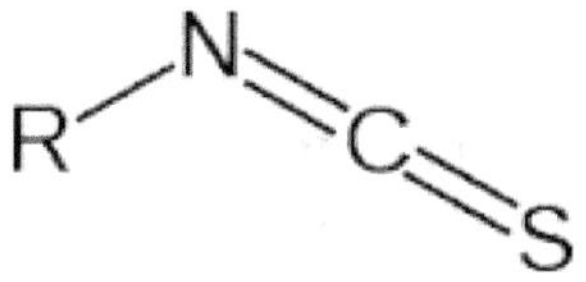</td></tr>
<tr><td>Mandelonitrile or cyanaogenetic
➢ Prunacin from wild cherry bark
➢ Amygdalin from bitter almond</td><td>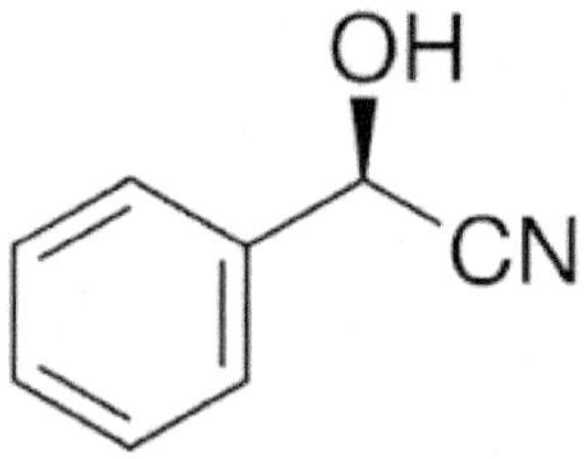</td></tr>
</table>

Contd...

Phenol ➤ Arbutin from bearberry	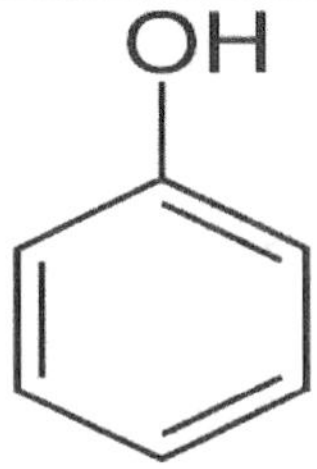 OH on a benzene ring (phenol structure)
Aldehyde ➤ Glucovanillin from vanilla pods	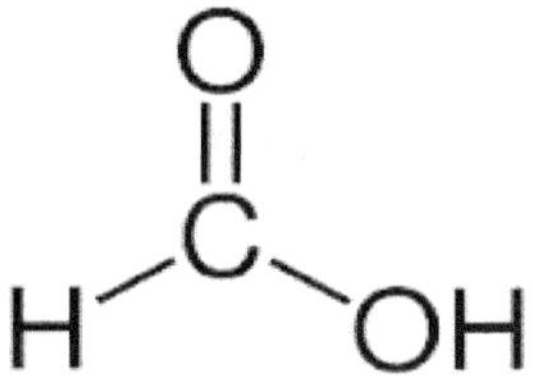 $H-\overset{\overset{\textstyle O}{\|}}{C}-OH$
Alcohol ➤ Salicin from *Salix* species	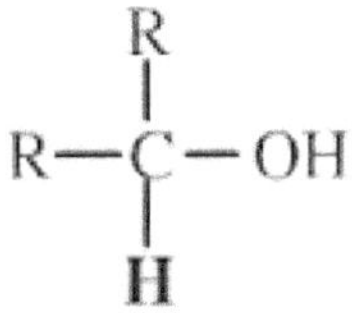 $R-\overset{\overset{\textstyle R}{\|}}{\underset{\underset{\textstyle H}{\|}}{C}}-OH$

General Method of Total Glycoside Extraction

The general method of total glycoside extraction involves "Stas-Otto" process. Stas-Otto method is a standard process developed in 1850. Since then, it has been subjected to a few minor changes over a period of time to overcome its shortcomings, but it still remains much the same. The standard Stas-Otto Method is as follows:

1. Take the accurately weighed and finely powdered sample.
2. Perform Soxhlet extractions of the sample using alcohol.
3. Collect the extract and treat it with a sufficient amount of lead acetate and filter it after sometime.
4. Pass H_2S gas through the filtrate and filter it again to remove the precipitates of lead sulphide.
5. Purify the filtrate by evaporation, fractional distillation, or by any chromatographic methods in order to obtain pure glycoside fraction.

Successive soxhlet extraction is done to remove the interfering substances by precipitation. It is a time consuming process and leads to loss of glycosides, hence the process can be optimised by applying vacuum. Glycosides are thermolabile in nature, hence the temperature of the soxhlet should be maintained below 45°C. Lead acetate is employed to precipitate tannins. But lead is poisonous; therefore, it is necessary to pass H_2S gas to remove it completely.

Antraquinone glycosides

Senna

Synonym: Sonamukhi, cassia senna, senna ki patti

Biological source: It consists of the dried leaflets of Alexandria senna (*Cassia acutifolia* Delile) or Tinnevelly senna (*Cassia angusti folia* Vahl) belonging to plant family legu-minosae

Geographical source: Tinnevelly senna is cultivated in South India especially Tinnevelly and nearby district of Tamilnadu state. Alexandrian senna is cultivated in Africa and Egypt region

Morphology: Following table gives details of morphological characters of both senna speices

Morphology	Cassia angustifolia (Tinnevelly or Indian senna)	Cassia acutifolia (Alexandrian senna)
Appearance	Less entire and more broken	Intire and less broken
Size	2-4 cm long, 7-12 mm wide	2.6-6 cm long, 7-8 mm wide
Shape	Ovate-lanceolate	Lanceolate
Margin	Intire and curled	Intire
Apex	Acute with sharp spine at the apex	Less acute with sharp spine at the apex
Base	More asymmetrical	Less asymmetrical
Texture	Thin and brittle	Firm and flexible
Surface	More pubescent	Less pubescent
Colour	Pale grayish-green	Pale green
Odour	Slight	Slight
Taste	Mucilagenous and slightly bitter	Mucilagenous and slightly bitter

Microscopy: Microscopic examination of the transverse section of senna leaflet reveals isobilateral structure. Paracytic stomata (**A**), nonlignified unicellular trichomes with warty walls (**B**) and fibrovascular bundle lined with abundant prisms and cluster crystals of calcium oxalate (**C**).

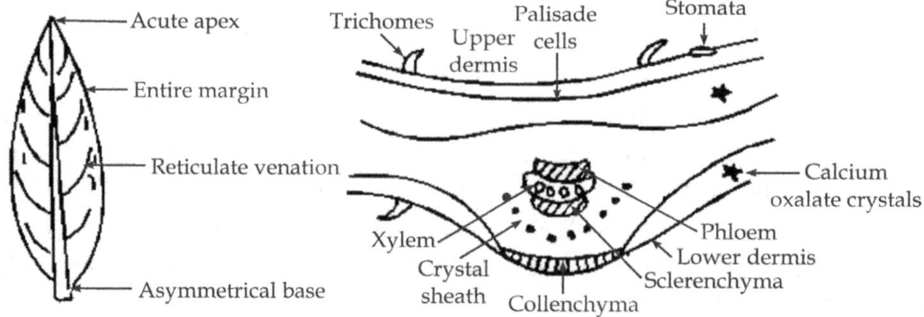

Cultivation

Method of propagation	Bysowing or drilling method
Time of propagation	In month Oct.-Nov. Feb-March
Soil	Red loamy or coarse gravelly soil pH 7.0-8.5.
Climate	dry summer with moderate temperature
1. Temperature	$30°$-$35°C$
2. Rainfall	25-40 cm
3. Altitude	600-900 m
Irrigation	Semi-irrigated crop, trough 5-6 irrigations.
Fertilizers	Nitrogenous fertilizers
Diseases	Leaf spots by Phyllostica sp. and dry rot by Macrophomina phaseoli
Pesticides	Dithane M-45
Harvestingtime and method	2-3 months after sowing, Harvest leaves in 3 stage by plucking
Drying and storage	Stored in bales under hydrolytic pressure, dry in shade
Yield	1200 kg/ha

Chemical constituents: The phytoconstituents principally respon-sible for purgative action are two anthrax-quinone glycosides namely; sennoside A and sennoside B. Sennoside A and B together are responsible for up to 40 – 60% activity of crude senna. Senna also contains small quantities of other anthraquinones such as sennosides C and D, rhein 8-glu-coside, aloe-emodin, 8-glucoside, anthronediglucoside and rhein. Additionally senna contains napthalene glycosides (tinnevellingly-coside and 6-hydroxy musizin glycoside), flavonoid (kaempferol), phytosterols, myricyl alcohol, salicylic acid, chrysophenic acid, mucilage, resin and calcium oxalate.

Sennoside A (Trans)   Homo-dianthrone   Sennoside B (Meso)   Homo-dianthrone   Sennoside C (Trans)   Hetero-dianthrone   Sennoside D (Meso)   Hetero-dianthrone

Chemical Tests

Borntrager's test	Take little quantity of aqueous solution of sample; add H_2SO_4, then add CCl_4 or ether. Separate the organic layer and shake with dilute ammonia	Rose pink colour of ammonia layer.
Schonteten's test for anthranols	Take little quantity of aqueous solution of sample and add sodium borate	Green fluorescence
P-nitrosodimethyl aniline test for Anthrones	Take little quantity of aqueous solution of sample and add P-nitrosodimethyl aniline	Anthrones changes colour of solution. Anthraquinone do not change colour.

Uses: Senna owes its characteristic pharma-cological effects primarily to the presence of hydroxyanthracene glycosides viz. senno-sides A and B. These glycosides increase secretion of gastric fluids and affect colonic motility and hence, facilitate colonic transit. These phytocontituents remain unabsorbed in the upper intestinal tract. They are activated by large intestine bacteria into the active derivatives (rhein-anthrone). Senna usually produces its characteristic action in 8-10 hours. Hence, its use at night is recommended. Sennosides relieve severe constipation. Chronic use; however; may cause alteration in water equilibrium and electrolyte metabolism and pigmentation of the intestinal mucosa.

Adulteration: The leaves ofthe following plants have also been reported as mixed with or substituted for senna

1. *Cassia auriculata* (Palthe senna): Leaves are small, oblong or obovate and conspicuously mucronate

2. *Cassia holoserica:* Leaves are smaller, more obtuse and hairy

3. *Cassia Montana*: Leaves are darker in colour, rounded apex and dark network of veins

4. *Solenostemma argel* (Argel leaves): Leaves are thick, rigid texture and peculiarly curled, curved, or twisted appearance, surface finely wrinkled, veins not evident, leaf equal at base and taste is distinctly bitter.

5. *Tephrosia apollinea*: Leaves are obovate-oblong in shape, pubescent, emargi-nated, lateral veins straight and parallel.

6. *Coluteaarborescens*(Bladder senna): Leaves are green and very thin.

7. *Ailanthus glandulosa*: Leaves are large, triangular-ovate in shape, strongly striated cuticle and no stomata on upper epidermis.

8. *Globularia alypum*(Province senna): Leaves are spathulate, mucronate, rounded apex and prism of calcium oxalate present in the epidermal cells.

9. *Coriariamyrtifolia*: Leaves are ovate-lanceolate in shape, grayish-green in colour, having two prominent lateral veins and conspicuous midrib.

10. *Cassia obovata*(Dog senna): Leaves broadly obovate, apex abruptly tapering, pinnate venation.

Aloe

Synonym: Mussabber, Kumari, Korpad,Ghritkumari

Biological Source: Aloes is the dried juice of the leaves offollowing plants belonging to Family Liliaceae

Aloe barbadensis, Aloe Vera	Curacao Aloe
Aloe ferox and Aloe Africana	Cape Aloe
Aloe perryi baker	Socotrine and Zanziber

Geographical source: Aloe is indigenous to Eastern and Southern Africa but it is also cultivated in Europe and many parts of India.

History: The word 'Aloe' is originated from an Arabic word 'Alloch' meaning shining bitter substance.

Morphologyand Microscopy:

Porperty	Curacao Aloe	Cape Aloe	Socotrine	Zanziber
Color	Yellow to chocolate brown	Greenish brown	Yellow to brown	Light blue
Odor	Iodoform like	Sour and distinct	Unpleasant	Characteristic
Taste	Bitter	Bitter	Bitter	Bitter
Texture	Opaque	Glassy	Opaque	Opaque
Fracture	Waxy	Brittle	Porous	Smooth
Microscopy of powder	Small needles	Transparent irregular fragments	Prisms	Lumps with irregular masses
Histological characters of leaf	Cuticle, epidermis, palisade cells, acicular bundles of calcium oxalate crystals, pericylicfibres, numerous stomata, vascular bundle, mucilaginous parenchyma, aloetic juice			

Chemical constituents: C – glycoside 'Aloin' is mixture of barbaloin, beta barbaloin and isobarbaloin. It also contains free aloe emodin. It contains resin such as, Aleosin-A and B. It also contains aloetic acid, galactouronic acid, choline, choline saliafate and saponin and mucopolysaccharides. Aloegel consists primarily of water and polysac-charides (pectins, hemicelluloses, glucomannan, acemannan and mannose derivatives). It also contains amino acids,   lipids, sterols (lupeol, campesterol and sitosterol, tannins, and enzymes. Mannose 6-phosphate is a major sugar component.

Barbaloin   Aloin A   Aloin B

Chemical Tests

Borax test	heat 5 ml solution with 0.2 gm borax; add few drops of this in a test tube filled with water	Green fluorescence is produced
Nitric acid test	Take 2.5 ml aqueous solution and add 2 ml nitric acid	Curacao aloe: Reddish brown color. Socotrine aloe: Pale yellow brown color Cape aloe: Brown changing to green.
Nitrous acid test	Take extract Solution and add Nitrous acid and acetic acid. heat	Reddish brown color
Modified Borntrager test for C-glycosides	Take extract solution, add $FeCl_3$ heat and filter. To filtrate add benzene. Separate benzene layer and strong ammonia solution	Lower ammonical layer shows pink to red color.

Uses: Aloe is used as purgative due to presence of C–glycosides. It is also anti inflammatory due to presence of salicylates and carboxypeptidase. It is one of the ingredients of compound tincture of benzoin. In cosmetics, it is used as protective, antiwrinkle, nourishing, cleansing and soothing ingredient. Aloe Gel increases removal of dead tissue due to presence of Aloctin which stimulates macrophage production. It is also used in shampoos and soaps.

Cyanogenetic Glycosides

Bitter Almond

Synonym: Amygdala amara,: Badam

Biological Source: It is dried ripe seed of plants *Prunus amygdalus* Batsch var amara belonging to family Rosaceae.

Description: Two major types of almonds are grown commercially, which can be categorized as sweet almonds (Prunus amygdalus dulcis) and bitter almonds (Prunus Amygdalus amara). The sweet almond producing plant and the bitter almond producing plant can be differentiated on the basis of their flowers, since the sweet almond flowers are white in colour, whereas the bitter almond flowers are pink in colour.

The bitter almond is slightly broader and shorter than the sweet almond and it contains about 50% of the fixed oil that occurs in sweet almonds. Bitter almonds yield 4-9 mg of hydrogen cyanide per almond.

Geographical Source: It is native to Iran and Asia and extensively cultivated in Italy, Spain, Portugal, France and **Morocco**.

Morphology:

Colour: Brown

Odour: Odourless

Taste: Bitter

Size: About 20 mm in length, 125 mm in width, 10 mm in thickness

Shape: Flat, oblong, ovoid with marking on testa

Chemical constituents: It contains fixed oil, protein, volatile oil majorly and small amount of cynogenetic glycoside amyg-dalin (2-4%). 'Amygdaline' differentiates the bitter almond from the sweet almond. In the presence of water (hydrolysis), amygdaline yields glucose and the chemi-cals, benzaldehyde and hydrocyanic acid (HCN). HCN, the salt of which is known as cyanide, is poisonous.

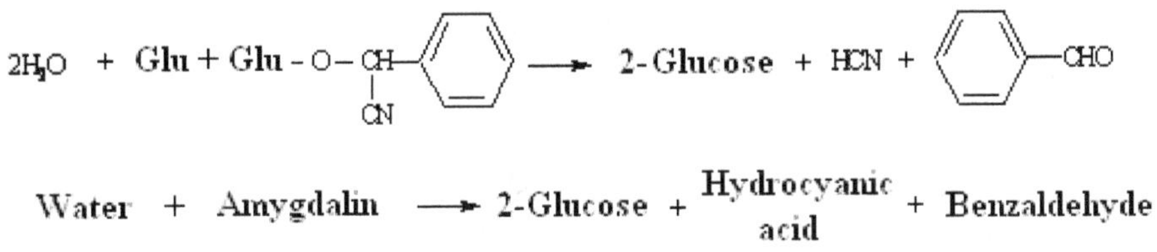

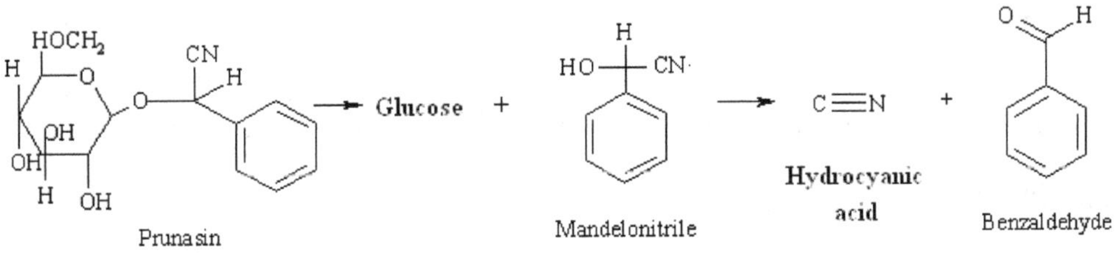

$$2H_2O + Glu + Glu-O-\underset{CN}{CH}-C_6H_5 \longrightarrow 2\text{-}Glucose + HCN + C_6H_5-CHO$$

$$Water + Amygdalin \longrightarrow 2\text{-}Glucose + Hydrocyanic\ acid + Benzaldehyde$$

Prunasin $\longrightarrow$ Glucose + Mandelonitrile $\longrightarrow$ Hydrocyanic acid + Benzaldehyde

Uses: Sedative, Demulcent, Emollient

2.8   Iridoids, Other Terpenoids andNaphthaquinones

Iridoids are a type of monoterpenoids in the general form of cyclopentanopyran, found in a wide variety of plants and some animals. They are biosynthetically derived from 8-oxogeranial. Iridoids are typically found in plants as glycosides, most often bound to glucose. The chemical structure is exemplified by iridomyrmecin, a defensive chemical produced by the ant genus

Iridomyrmex, for which iridoids are named. Structurally, they are bicyclic cis-fused cyclopentane-pyrans. Cleavage of a bond in the cyclopentane ring gives rise to a subclass known as secoiridoids, such as oleuropein and amarogentin.

Iridoid glycosides are a large class of natural products prevalent in the plant kingdom. Derived from the monterpenoidiridotrial, they can be broadly divided into two subclasses: iridoids, containing an intact cyclopentene ring, and secoiridoid

Iridoids

Iridoids contain iridane skeleton which is actually a monoterpenoid but different in type of folding of cyclopentane ring fused to a six-membered oxygen heterocycle. Glycosides of iridoids are called as loganins (Example: nepetalactone). Loganin is a key intermediate in the biosynthesis of many other iridoid structures, and also features in the pathway to a range of complex terpenoid indole and tetrahydroisoquinoline alkaloids. Example: strychnine, yohimbine, vinca alkaloids, and ellipticine. Cleavage of the simple monoterpene loganin skeleton gives secologanins, secoiridoids (Example: Gentiopicroside and epoxyiridoids (Example: valepotriates). Iridoids are mainly found in families Ericaceae, Loganiaceae, Gentianaceae, Rubiaceae, Verbenaceae, Lamiaceae, Oleaceae, Plantaginaceae, Scrophulariaceae, Valerianaceae, and Menyanthaceae.

Iridoid   Iridoid glycoside   Seco-iridoid   Seco-iridoid glycoside

Naphthoquinones

Quinones are commonly-distributed aromatic compounds present in plants as well as fungi, lichens, algae and bacteria. Based on chemical structure, quinones are classified into benzoquinones, anthraquinones and naphthoquinones. Naphthoquinones, structurally related to naphthalene, are natural pigments whose colors range from yellow to red. Vitamin K1 is most common napthaquinone. Naphthoquinones are found to exhibit cytotoxic, antimicrobial, antifungal, antiviral and antiparasitic properties. Some napthaquinones are polyketides like plumbagin and 7-methyljuglone and some are chorismic acid derivative like lawsone and juglone.

Lawsone Source: Leaves and stems from henna (Lawsoniainermis L.) 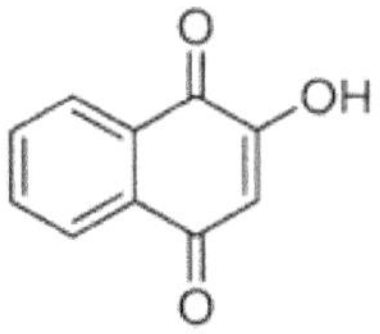	**Naphthazarin** Source: Wood bark tree Lomatia obliqua and Alkana species 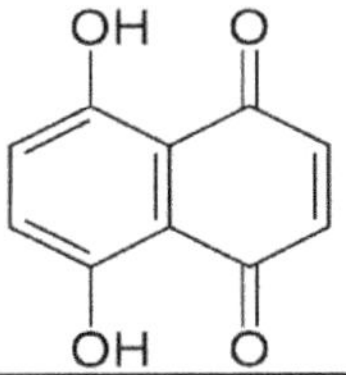
Plumbagin Source: Roots of Plumbago scandens 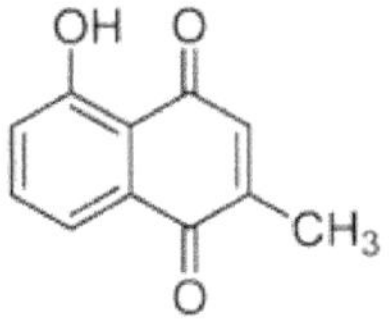	**Mompain** Source: Fungi Helicobasidiummompa 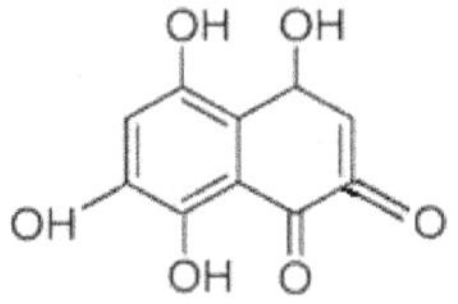
Lapachole Source: Heartwood of plants of the genus Tabebuia spp., Tecoma spp. and Tecomellaundulata (Chemically, it is a derivative of vitamin K) 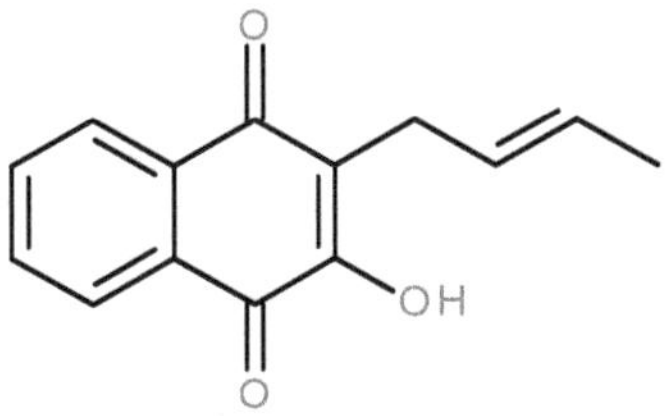	**Shikone** Source: Roots of the plant Lithospermum erythrorhizon 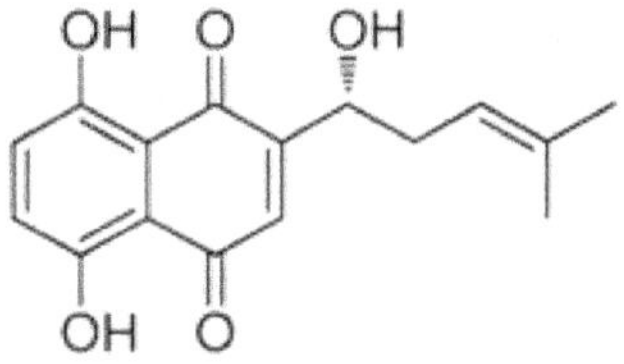
Juglone Source: Roots, leaves, nuts, bark and wood of black walnut (Juglans nigra), European walnut (Juglans regia) and American white walnut (Juglans cinerea) 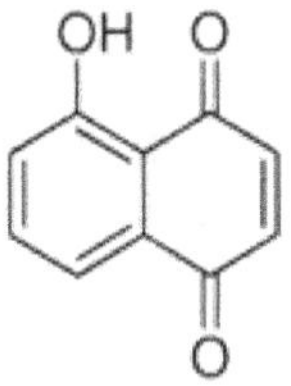	**Shikonin** Source :Roots of Alkanna tinctoria (Enantiomer of shikone) 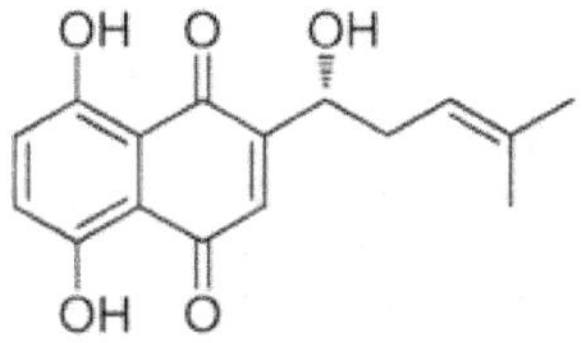

Gentian

Synonym: gentian root, bitter root

Biological Source: It is dried rhizome and root of plant *Gentiana lutea*, Family Gentianaceae

Geographical source: Hill areas in southern and central Europe like Jura, Vosges Mountains and Yugoslavia.

Macroscopy:

Colour: Yellowish brown

Odour: characteristic, agreeable

Taste: first sweet followed by intensely bitter

Extra feature: fracture is short, smooth in dried drug but tough flexible in drug.

Microscopy: It shows bark, cambium, wood and pith. Root shows primary xylem, cork cells are thin walled, cortex has parenchyma with oil globules and calcium oxalate, phloem is present in small group and phloem fibers are absent.

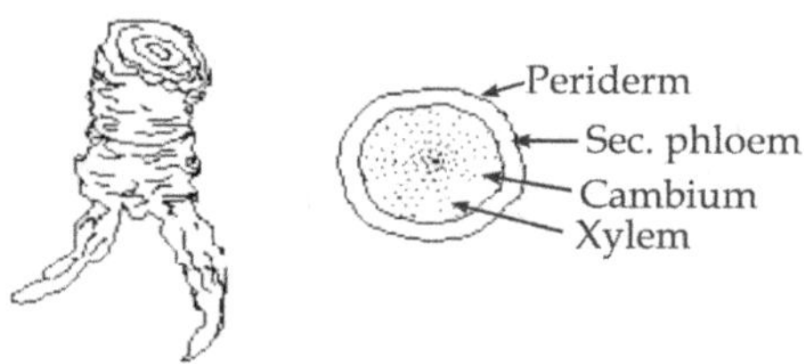

Chemical Tests: Gentian extract shows light blue fluorescence's when observed under UV.

Chemical constituents: It contains Bitter glycoside: gentiopicrin (Gentiopicroside), flavonoid alkaloid gentisic acid, gentianose, amarogenin (strongly bitter glucoside), amaroswerin, gentioside; gentinin (mixture of gentiopicrin and gentinin).

Uses: It is used as bitter tonic to in anorexia and dyspepsia to stimulate gastric secretion hence improving appetite

Artemisia

Common name: Sweet wormwood, sweet annie, sweet sagewort, annual mugwort, annual wormwood

Biological source: Leaves of plant *Artemisia annua* belongs to the family of Asteraceae

Geographical source: native to temperate Asia, but naturalized in many countries including scattered parts of North America

Morphology:Annual plant, aromatic, green, glabrous or with scattered, small, approximate hairs. Stem erect, ribbed, brownish or violet-brown, naturally grows to 30–100 cm high (cultivated plants may reach 200 cm high). Leaves alveolate-punctateglandular; lower leaves petiolate, 3–5 cm long and 2–4 cm wide, ovate, thrice pinnatley cut, their lobules oblong-lanceolate, short-acuminate, entire or with 1–2 teeth, 1–2 mm long and 0.5 mm wide; middle and cauline leaves twice pinnately cut; upper leaves sessile, smaller and less compound; uppermost leaves bracteal, simple with fewer lateral lobes

Colour: ranges from green to brown-green.

Odour: crushed leaves have a characteristic odour.

Taste: crushed leaves have a slightly bitter taste.

Microscopy: The leaves are dorsiventral histologicallly.Spongy parenchyma contains 4-6 layers of loosely arranged cells. Reticulate xylems are lignified and present in the ventral surface of the leaf. The leaves contain anomocytic stomata with numerous glandular and non-glandular trichomes on both the surfaces with little stalk.

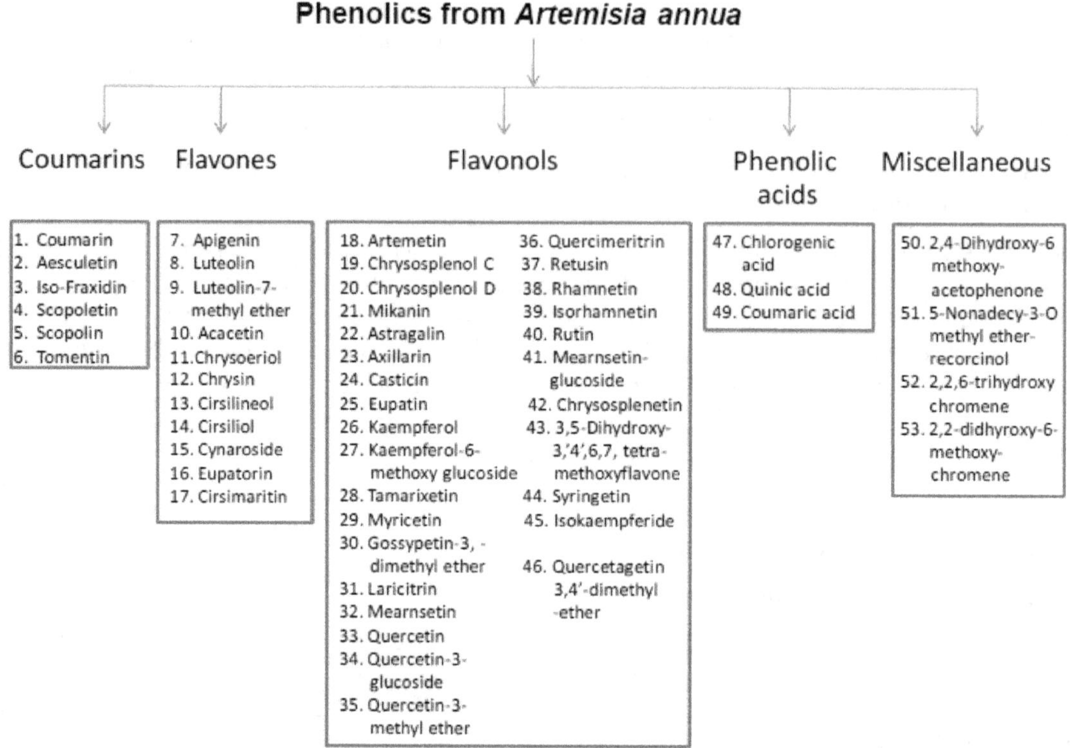

Chemical constituents: The chemical composition of Artemisia annua consists of volatile and non-volatile constituents. The volatile components are mainly attributable to essential oils with the content of the latter being 0.2–0.25%. The main compounds, which account for about 70% of the essential oils, appear to be camphene, β-camphene, isoartemisia ketone, 1-camphor, β-caryophyllene, β-pinene, artemisia ketone, 1, 8-cineole, camphene hydrate, and cuminal are also found.

Five major polyphenolic groups (coumarins, flavones, flavonols, phenolic acids and miscellaneous) containing over 50 different phenolic compounds are also identified. The main non-volatile ingredients include sesquiterpenoids, flavonoids and coumarins, together with proteins(suchas β-galactosidase, β-glucosidase), steroids (e.g.Beta-sitosterol and stigmasterol).

The main chemical constituents of *Artemisia annua* are sesquiterpenoids, including artemisinin, artemisinin I, artemisinin II, artemisinin III, artemisinin IV, artemisinin V, artemisic acid, artemisilactone, artemisinol and epoxyarteannuinic acid.

artemisinin

dihydroartemisinin

artesunate

artemether

arteether

artemisone

Uses: In traditional Chinese medicine, *A. annua* is traditionally used to treat fever and thus malaria. The proposed mechanism of action of artemisinin involves cleavage of endoperoxide bridges by iron, producing free radicals which damage biological macromolecules causing oxidative stress in the cells of the parasite. It appears to be a much targeted oxidative agent, and one that is not cyclically oxidized and reduced.

Taxol

Common name: English yew or European yew

Biological source: Bark, leaves and roots of *Taxus* species like *T.baccata* (European or English yew), *T. brevifolia* (Pacific yew or Western yew), *T. canadensis* (Canadian yew), *T. chinensis* (Chinese yew), *T.cuspidata* (Japanese yew), *T. floridana* (Florida yew), *T. globosa* (Mexican yew) and *T. wallichiana* (Himalayan yew) of family Taxaceae. In 1993, taxol was discovered as a natural product in a newly described endophytic fungus living in the yew tree. It has since been reported in a number of other endophytic fungi.

Geographical source: Pacific yew native to the Pacific Northwest of North America. Common names of most of taxus species are from their geographical source.

Morphology: Odorless, bitter, red aril and seed cones, needle shaped leaves, reddish brown bark, yellow flowers

Microscopy (Pacific Leaves Yew):

The upper section shows elliptical shaped epidermal cells. The lower section shows a marginal region of eight (8) smooth trapezoidal to almost rectangular cells in width with no (0) transitional zone of papillose cells; this is followed by a stomata band with 14 rows of stomata, a midrib of 15-18 cells in width that lack papillae, and part of second stomata band

Bark: Tracheid pits in the radial walls generally uniseriate. Rays homocellular, without ray tracheids. Ray pits cupressoid. Transversal walls of rays more ore less thick, tangential walls of ray cells thin, indentures present at junction of transverse and tangential walls. Conspicuous spiral thickenings in longitudinal tracheids.

Chemical constituents: Main constituents are paclitaxel/taxol and its semisynthetic derivatives-10-deacetyl baccatin (docetaxel/ taxotere), larotaxel, cabazitaxel, TPI287/NBT-287.

Paclitaxel (TaxolTM) R₁ = Ph, R₂ = COCH₃, R₃ = H
Docetaxel R₁ = OC(CH₃)₃, R₂ = R₃ = H
Cabazitaxel R₁ = OC(CH₃)₃, R₂ = R₃ = CH₃

Larotaxel

Uses: Taxane compounds are used in ovarian cancer, breast cancer, lung cancer, Kaposi sarcoma, cervical cancer, and pancreatic cancer treatment. It works by interference with the normal function of microtubules during cell division. Paclitaxel ((Taxol®, Bristol-Meyers Squibb), stabilizes the microtubule polymer and protects it from disassembly. Chromosomes are thus unable to achieve a metaphase spindle configuration. This blocks the progression of mitosis and prolonged activation of the mitotic checkpoint triggers apoptosis or reversion to the G-phase of the cell cycle without cell division.

Paclitaxel and its semi-synthetic analog docetaxel (Taxotere®, Sanofi-Aventis) have become a mainstay in the treatment of solid tumors, including breast and ovarian cancer. Although the taxanes have shared clinical success for many years, serious limitations of these drugs include the dose-limiting toxicities of immunosuppression and peripheral neuropathy as well as inherent and acquired drug resistance. The therapeutic utility of taxanes, the most widely used class of anticancer agents, is hampered by resistance due to over expression of MDR-1 and tubulin mutations. The most well established, clinically relevant form of taxane resistance is overexpression of the P-glycoprotein ATP-binding cassette (ABC) drug transporter (Pgp). Intrinsic overexpression of Pgp in the liver, kidney and intestinal tract has limited the use of the taxanes and other Pgp substrates in tumors derived from those tissues.

One serious problem in the development of the taxanes is their poor solubility. This property necessitated their formulation in cremophor, an agent that causes hypersensitivity reactions and requires patient pretreatment. Abraxane® (Albumin-bound Paclitaxel for Injectable Suspension) is a paclitaxel derivative that has an increased intrinsic solubility conferred by the conjugation of albumin to paclitaxel, eliminating the requirement for cremophor. The increased solubility of

Abraxane® dramatically decreases the time required for drug administration from 3 h to 30 min. The dimethyl ether paclitaxel analogue cabazitaxel was approved in 2010. One of its advantages is that it is less susceptible to drug resistance than paclitaxel and docetaxel.

Subjective Questions

1. What is irridoid? Explain with examples.

2. What is difference between Digitalis lanata and digitalis purpurea?

3. What is difference between Indian senna and Alexandrian senna?

4. Why does Borntrager's test modified?

5. What are different types of aloe and how to differentiate between these aloes?

6. How to differentiate sweet almond vs bitter almond?

7. What is lignan? Explain with examples.

8. Write a note on Belladonna or Digitalis

9. Give biological source, description, chemical constituents, chemical tests and uses

    a. Clove   b. Taxol   c. Ruta   d. Myrrh

10. Give biological source, description, chemical constituents, chemical tests and uses

    a. Catechu   b. Bitter Almond   c. Artemisia   d. Pterocarpus

11. Differentiate Guggul and Myrrh.

12. Name the semi synthetic derivatives of Artemisinin, podophyllotoxin and paclitaxel.

13. List side effects of licorice, digitalis and rauwolfia.

14. Differentiate senna, digitalis, belladonna and tea leaves based on morphological and microscopical characters.

15. What is difference between mode of action of vinca, taxol and podophyllotoxin.

16. What are the benefits of semisynthetic derivatives of paclitaxel?

17. What is katha and Kutch?

Multiple Choice Questions (MCQs)

1. Periwinkle is synonym of ……..
   a. Catharanthus roseus
   b. Physostigmavenenosum
   c. Rauwolfia serpentine
   d. Papaver somniferum Linn.

2. *Vinca rosea* is indigenous to……..
   a. South Africa
   b. Madagascar
   c. India
   d. Indonesia

3. In which country Vinca was used previously for toothache?
    a. Brazil
    b. South Africa
    c. Jamaica
    d. Australia

4. In vinca, cruciferous stomata are present more frequently on..........
    a. Upper epidermis
    c. Lower epidermis
    b. On both sides of leaf
    d. None

5. Which one of the following is NOT active constituent of *Catharanthus roseus*?
    a. Vincristine
    b. Vinblastine
    c. Serpentine
    d. Rescinnamine

6. Chhotachand is synonym of .... drug?
    a. Catharanthus roseus
    c. Rauwolfia serpentine
    c. Physostigmavenenosum
    d. Papaver somniferum Linn.

7. Reserpine was first isolated by ....in 1952?
    a. Muller
    b. White
    c. Galen
    d. None

8. For healthy plant growth of *Rauwolfia serpentina*, soil should NOT contain........
    a. Large amount of water
    b. Large amount of sand
    c. Large amount of clay
    d. None

9. Total alkaloid content of rauwolfia root ranges from..............
    a. $0.7 - 3\%$
    b. $7 - 3\%$
    c. $7 - 0.3\%$
    d. $5 - 0.3\%$

10. Syrosingopine is also called ....
    a. Methyl carbethoxysyringoylreserpate
    b. Ethyl carbethoxysyringoylreserpate
    c. Propyl carbethoxysyringoylreserpate
    d. None

11. Reserpine shows ....colour when treated with solution of vanillinin acetic acid
    a. Violet blue
    c. Violet red
    c. violet yellow
    d. Violet orange

12. Reserpine lowers the blood pressure by ....
    a. Enhancing stores of catecholamines
    b. Depleting stores of catecholamines
    c. Inhibiting stores of catecholamines
    d. Both a and b

13. Recommended daily dose ofRauwolfia is.........
    a. 100 – 150 mg (oral twice daily)
    b. 200 – 250 mg (oral twice daily)
    c. 300 – 350 mg (oral twice daily)
    d. 10 – 50 mg (oral twice daily)

14. Which one of the alkaloid isolated from rauwolfia causes mental depression in higher doses?
    a. Reserpine
    b. Ajmalicine
    c. Rescinnamine
    d. Deserepidine

15. In India, all activities related to opium are controlled under ....
    a. Narcotic Drugs and Psychotropic Substances Act 1995
    b. Narcotic Drugs and Psychotropic Substances Act 1985
    c. Narcotic Drugs and Psychotropic Substances Act 1975
    d. Narcotic Drugs and Psychotropic Substances Act 1965

16. Which variety of Opium is cultivated in India?
    a. P. somniferum var. glabrum
    b. P. somniferum var. album
    c. P. somniferum var. nigrum
    d. None

17. The instrument used to collect exuded latex is called ....
    a. Tripala
    b. Favada
    c. Charpala
    d. Khora

18. Where the Government opium processing factory is situated in India?
    a. Ghaziabad
    b. Ghazipur
    c. Burhanpur
    d. Hariyana

19. Narcotine, Narceine and papaverine belongs to .... alkaloid category.
    a. Benzyl-isoquinoline
    b. Phenanthrene
    c. Indole
    d. None

20. The opium alkaloids are present as ....
    a. Salts of Meconic acid
    b. Salts of Nitric acid
    c. Salts of Mevalonic acid
    d. None

21. Which one of the following is an important allied plant of opium which does not contain morphine as an addictive substance?
    a. P. intermedia
    b. P. paeoniflorum
    c. P. bracteatum
    d. P. orientate

22. Deadly night shade leaf is synonym of …..
    a. Indian belladonna
    b. European belladonna
    c. Indian senna
    d. European senna

23. Biological source of Indian belladonna is………
    a. Atropa belladonna Linn.
    b. Atropa acuminata Royle ex – Lindley
    c. Solanum nigrum
    d. None of the above

24. Palisade ratio of Atropa belladonna is …………..
    a. 4-7
    b. 6-7
    c. 5-7
    d. 7-8

25. Which one of the following drug is used as an antidote in opium and chloral hydrate poisoning?
    a. Rauwolfia serpentine
    b. Chlorophytum borivilianum
    c. Atropa belladonna
    d. Azadirachta indica

Phenylpropanoids and Flavonoids

26. Phenylpropanoids organic compounds synthesized by plants from amino acids……….
    a. Tryptophan and alanine
    b. Phenylalanine and tyrosine
    c. Tyrosine and tryptophan
    d. Phenylalanine and tryptophan

27. The term Lignan was first introduced in 1948 by ……………………….to describe group of phenylpropanoids
    a. Gotleib
    b. Haworth
    c. Umezawa
    d. None of the above

28. Chemically lignans are………
    a. Two $C_6 - C_3$ units attached by its 8th central carbon atom
    b. Two $C_6 - C_3$ units attached by its 9th central carbon atom
    c. Two $C_6 - C_3$ units attached by its 10th central carbon atom
    d. Two $C_6 - C_3$ units attached by its 6th central carbon atom

29. The biological source of Lignans, Sesamin and sesamolin is …………..
    a. Carthamus tinctorius L
    b. Linum usitatissimum L
    c. Hydnocarpus heterophylla
    d. Sesamum indicum

30. Lignans exerts potential………..
    a. Anti-inflammatory activity
    b. Antioxidant activity
    c. Anticancer activity
    d. All of the above

31. In which form flavonoid compounds are present in plants?
    a. Alkaloids
    b. Glycosides
    c. Tannins
    d. Lipids

32. Which one of the classes of flavonoids consists of water soluble plant pigments?
    a. Anthoxanthin
    b. Anthraquinones
    c. Anthocyanin
    d. Both a and b

33. The class of flavonoids which represents flavones is known as ……..
    a. Anthocyanin
    b. Anthoxanthin
    c. Both a and b
    d. None of the above

34. Which one of the following is biological source of rutin?
    a. Fagopyrum esculantum
    b. Gingko biloba
    c. Sylibum marianum
    d. Ammi visnaga

35. Which one of the following drug is used in treatment of retinal hemorrhages?
    a. Rutin
    b. Quercetin
    c. Khellin
    d. Visnagin

36. Buck wheat is synonym of
    a. Sophera japonica
    b. Fagopyrum esculantum
    c. Both a and b
    d. None of the above

37. Rutin is …………….
    a. Rhamnoglycoside of quercetin
    b. Rhamnoglucoside of quercetin
    c. Rhamnoglycosideof hesperidin
    d. Rhamnoglucosideof hesperidin

38. Camellia thea is synonym of
    a. Coffee
    b. Cocoa
    c. Tea
    d. Kola

39. Thea sinensis belongs to family……..
    a. Theaceae
    b. Ternstroemiaceae
    c. Both a and b
    d. None

40. Green tea originated from ……….
    a. India and Sri Lanka
    b. China and Japan
    c. Korea
    d. America and Russia

41. Green tea is synthesized by ………..
    a. Putting tea leaves in iron pans and drying by artificial heat
    b. Putting tea leaves in iron pans and drying by sunlight
    c. Putting tea leaves in copper pans and drying by artificial heat
    d. Putting tea leaves in copper pans and drying by sunlight

42. The colour of the tea leaves is due to ………..
    a. Theobromine
    b. Caffeine
    c. theophyline
    d. gallotannic acid

43. Blacktea contains relatively low amount of polyphenolic compound epigallacto catechin gallate as compared to green tea because of -
    a. Fermentation process
    b. Artificial drying
    c. drying in sunlight
    d. None

44. Which type of tea is rich source of flavonoids and polyphenolic compounds?
    a. Black tea
    b. Green tea
    c. Both a and b
    d. None

45. How much amount of caffeine is present in tea leaves?
    a. 1-3%
    b. 2-4%
    c. 4-5%
    d. 1-2%

Steroids, Cardiac Glycosides & Tri-terpenoids

46. Which variety of licorice is known as 'Spanish liquorice'?
    a. G. glbra var. glandulifera
    b. G. glbra var. typical
    c. G. glbra var. violacea
    d. None of the above

47. Glycyrrhizin is ………
    a. Potassium and calcium salt of glycyrrhizinic acid
    b. Potassium and calcium salt of glycyrrhizic acid
    c. Potassium and calcium salt of glycyrrhetinic acid
    d. Potassium and calcium salt of glycyrrhetic acid

48. Which variety of *Glycerrhiza glabra* contains highest amount of glycyrrhizin?
    a. Russian liquorice
    b. Persian liquorice
    c. Spanish liquorice
    d. Both a and c

49. Which one of the following compounds prepared from glycerrhiza possesses significant minerocorticoid activity and used as anti ulcer drug?
    a. Liquiritin
    b. Isoliquiritin
    c. Carbenoxolone
    d. None

50. Glycyrrhizin is employed in the treatment of ....
    a. Rheumatoid arthritis
    b. Addison's disease
    c. Both a and b
    d. None of the above

51. Manchurian liquorice is obtained from ....
    a. Glycyrhizza glabra
    b. Glycyrhhizauralensis
    c. Glycyrrhiza glandulifera
    d. None

52. Which one of the following is known as rheumatism root?
    a. Rauwolfia serpentine
    b. Dioscoreadeltoidea
    c. Chlorophytum borivilianum
    d. Glycerrhiza glabra

53. Diosgenin is the hydrolytic product of ....
    a. Smilagenin
    b. Epismilagenin
    c. yammogenin
    d. dioscin

54. Dioscorea is used mainly in treatment of
    a. Psoriasis
    b. Rheumatoid arthritis
    c. epilepsy
    d. Ulcer

55. Which one of the following is alternative potential source of diosgenin other than Dioscorea species?
    a. Costusspeciosus
    b. Chlorophytum borivillianum
    c. Dioscorea composita
    d. None

56. Dioscorea generally grows at an altitude of ....
    a. 2000 – 3000 m
    b. 1000 – 2000 m
    c. 1000 – 3000 m
    d. 3000 – 4000 m

57. Which one of the following compound is used as precursor for corticosteroids?
    a. 16 – dehydropregnenelone acetate
    b. 16 – dehydropregnenone acetate
    c. 16 – dehydroprgesterone acetate
    d. None

58. Discoreaflouribunda contains ....
    a. 7 % of diosgenin
    b. 10 % of diosgenin
    c. 5% of diosgenin
    d. 8% of diosgenin

59. In case of digitalis, quality of secondary metabolites highly depends on ……………
    a. Temperature
    c. Sunlight
    b. Moisture content
    d. all of the above

60. Digitalis grows luxuriously at an altitude………..
    a. 1600 – 3000 m
    c. 1500 – 2000 m
    b. 1700 – 2500 m
    d. 1800 – 2500 m

61. Digitalis leaves should NOT contain moisture …………..
    a. More than 10%
    c. More than 5%
    b. Less than 10 %
    d. Less than 5%

62. Which one of the following is an important characteristic of Digitalis?
    a. Glandular Trichome
    c. Calcium oxalate crystals
    b. Collapsed celled covering Trichome
    d. starch grain

63. In digitalis, cardiac glycosides are present in quantity………..
    a. 0.2 – 0.45 per cent
    c. 0.2 – 0.55 per cent
    b. 0.2 – 0.35 per cent
    d. 0.2 – 0.65 per cent

64. Digitoxin is hydrolysis product of …………..
    a. Purpurea glycoside A
    c. Purpurea glycoside C
    b. Perpurea glycoside B
    d. Purpurea glycoside D

65. Which one of the following test is for identification of digitoxose?
    a. Baljet test
    c. Keller killiani test
    b. Legal test
    d. All of the above

Volatile Oils

66. ……….. part mentha oil is isolated from *Mentha piperata* Linn?
    a. Fresh leaves
    c. roots
    b. Fresh flowering tops
    d. Dry leaves

67. For good quality of mentha oil, *Mentha piperita* should be cultivated at temperatures between
    a. 25- 30°C
    c. 15 - 25°C
    b. 15 – 20 °C
    d. 30 - 40°C

68. The average oil content of mentha piperita is…….
    a. 1-2% (v/w)
    c. 1 - 2%(w/w)
    b. 0.5 – 1% (v/w)
    d. 0.5 – 1% (w/w)

69. Specific gravity of mentha oil is........
    a. 0.9 – 0.912g
    b. 1g
    c. 0.999 g
    d. 0.8 – 0.9 g

70. Which variety of Mentha is cultivated commercially in India?
    a. Japanese Mint
    b. American Mint
    c. Indian Mint
    d. None

71. Which component of mentha leave exerts anti-inflammatory and anti ulcer activity?
    a. Menthol
    b. Chavicol
    c. Menthone
    d. Azulene

72. Carryophyllum is synonym of ....
    a. Carum caraway
    b. Eugenia caryophyllum
    c. Santalum album
    d. None

73. Clove oil specific gravity is.........
    a. Greater than water
    b. Lighter than water
    c. equal to water
    d. None

74. Clove oil when treated with potassium hydroxide produces............
    a. Needle shaped crystals of potassium eugenate
    b. No reaction occurs
    c. White fumes
    d. None

75. Clove oil is used in ....
    a. Manufacture of vanillin
    b. Manufacture of cigarettes
    c. Both a and b
    d. None

76. Biological source of cinnamon is.........
    a. Bark of Cinnamomum verum
    b. Cinnamomum zeylanicum
    c. Both a and b
    d. None

77. Cinnamon belongs to family........
    a. Labiateae
    b. Lauraceae
    c. Myristicaceae
    d. Rutaceae

78. Cinnamon grows favorably at an altitude ....
    a. 2000 – 3000 m
    b. 800 – 1000 m
    c. 1000 – 2000 m
    d. 500 – 800 m

79. During manufacturing of cinnamon, small pieces and debris produced during handling of quills of cinnamon are called ……..
   a. Quills
   b. Oil
   c. Chips
   d. Double quill

80. Chief active constituent of cinnamon oil is ………
   a. Benzaldehyde
   b. Eugenol
   c. Cinnamaldehyde
   d. Cuminaldehyde

81. Chief constituent of coriander oil is………
   a. L – borneol
   b. Geraneol
   c. D- linalool
   d. Pinene

82. Optical rotation of coriander oil is …..
   a. +8° to + 15°
   b. +10° to + 15°
   c. +9° to + 15°
   d. +7° to + 15°

83. Biological source of fennel is……….
   a. Ripe fruits of *Coriandrum sativum*
   b. Ripe fruits of *Anethum graveolens*
   c. Ripe fruits of *Foeniculum vulgare*
   d. Ripe fruits of *Citrus limonis*

84. Anethole is chief active constituent……………..
   a. *Foeniculum vulgare*
   b. *Coriandrum sativum*
   c. Anethum graveolens
   d. None

85. Refractive index of fennel oil is ………….
   a. 1.526 – 1.538
   b. 1.108 – 1.124
   c. 1.342 – 1.1355
   d. 1.421 – 1.432

Tannins

86. Biological source of catechu is……….
   a. Dried aqucous extract of *Acacia catechu* wild
   b. Dried aqueous extract *Acacia chundra* wild
   c. Both a and b
   d. None

87. During extraction process of catechu, active constituents are divided in to two parts ……
   a. Katha
   b. Cutch
   c. Both a and b
   d. None

88. Catechin is present in ……
    a. Black catechu
    b. Pale catechu
    c. both and b
    d. None

89. Gambier fluorescein is present in …….
    a. Black catechu
    b. Pale catechu
    c. both and b
    d. None

90. Biological source of pale catechu is …….
    a. Dried aqueous extract of *Acacia catechu*
    b. Dried aqueous extract of *Acacia chundra*
    c. Dried aqueous extract of Uncaria gambier
    d. None

91. Chlorophyll pigments are ABSENT in ……
    a. Plae catechu
    b. Black catechu
    c. Both a and b
    d. None

92. Which one of the following is not present in Black catechu?
    a. Catechin
    b. Quercetin
    c. fluorescein
    d. Both a and b

93. Which part of the catechu extract has cooling and digestive properties?
    a. Katha
    b. Cutch
    c. Both a and b
    d. None

94. Black catechu belongs to family…………………
    a. Leguminosae
    b. Rubiaceae
    c. Scrophulariaceae
    d. Apocynaceae

95. Pale catachu belongs to family……………………..
    a. Leguminosae
    b. Rubiaceae
    c. Scrophulariaceae
    d. Apocynaceae

96. Synonym ofpterocarpus is ……
    a. Bijasal
    b. Malbar kino
    c. Indian kino tree
    d. All of the above

97. Biological source of Pterocarpus is ………
    a. Dried juice of *Pterocarpus marsupium*c. dried bark of Pterocarpus marsupium
    b. Dried leaves of Pterocarpus marsupium   d. Dried fruits of Pterocarpus marsupium

98. The chief constituent of Pterocarpus is ………
    a. Gallic acid
    b. K – pyrocatechin
    c. Kinotannic acid
    d. Resin

99. Kino red is ....
    a. Glucosidal tannin
    b. Anhydride of kinoin
    c. Glycosidal tannin
    d. Hydried of kinonin

100. Kinotannic acid is ....
    a. Glucosidal tannin
    b. Anhydride of kinoin
    c. Glycosidal tannin
    d. Hydried of kinonin

101. When the solution of Pterocarpus is treated with ferrous sulphate, it produces
    a. Red colour
    b. Violet colour
    c. Green colour
    d. Orange colour

102. When the solution of Pterocarpus is treated with alkali, it gives
    a. Violet colour
    b. Red colour
    c. Green colour
    d. Yellow colour

103. Kino exerts which one of the following therapeutic activity?
    a. Anti diabetic
    b. Astringent
    c. Antidiarrhoeal
    d. All of the above

104. Cutch obtained from catechu is used for ....
    a. Manufacture of ionexchange resin
    b. protective agent for fishing nets
    c. Water softening
    d. All of above

105. Standard ash value of black catechu is ....
    a. Not more than 10 per cent
    b. Not more than 6 per cent
    c. Not less than 10 per cent
    d. Not less than 6 per cent

Resins

106. Sumatra benzoin is ....
    a. Styrax benzoin
    b. Styrax paralleloneurus
    c. Styrax tonkinesis
    d. Both a and b

107. Siam benzoin is ....
    a. Styrax benzoin
    b. Styrax paralleloneurus
    c. Styrax tonkinesis
    d. Both a and b

108. The major constituent of siam benzoin is
    a. Vanillin
    b. Styrol
    c. Coniferyl benzoate
    d. Coneferyl benzoate

109. Biological source of Guggul is ....
    a. Stem bark of commiphoraweightii
    b. Stem bark of commiphoramolmol
    c. both a and b
    d. None of the above

110. 'Guggulip' developed from ....is an antihyperlipidaemic product
    a. Commiphoraweightii
    b. Commiphoramolmol
    c. Commiphoramukul
    d. None of the above

111. Ash value of Guggul should be ....
    a. Not more than 5%
    b. Not more than 3%
    c. Not more than 5.5 %
    d. Not more than 4.5 %

112. Gengerin is synonym of ....
    a. Rhizhomes of *Zingiber officinale*
    b. Oleo –Resin of *Zingiber officinale*
    c. Whole plant of *Zingiber officinale*
    d.None

113. Which one of the following is main pungent principle of Ginger?
    a. Zingerone
    b. Shogaol
    c. Gingerol
    d. All of the above

114. Shogaol is
    a. Pungent principle of ginger
    b. Non pungent principle of ginger
    c. Pungent principle of curcumin
    d. Non pungent principle of curcumin

115. Biological source of myrrh is oleo gum resin obtained from
    a. Commiphoramukul
    b. Commiphoraweighttii
    c. Commiphoramolmol
    d. None

116. What is the common characteristic of plant family Burseraceae?
    a. Plants posses oleo resinous canals in their conducting tissue
    b. Plants posses oleo resinous canals in their roots
    c. Plants posses oleo resinous canals in their rhizomes
    d. Plants posses oleo resinous canals in their flowering tops

117. Which one of the following is ether insoluble acid present in myrrh resin?
    a. α,β,γ commiphoric acid
    b. α,β,γ herrabomyrrholic acid
    c. α and β herrabomyrrholic acid
    d. α and β commiphoric acid

118. Which one of the following is identification test for myrrh?
    a. When triturated with water, produces yellowish – brown emulsion
    b. When triturated with water, produces brownish – yellow emulsion
    c. When triturated with water, produces reddish– brown emulsion
    d. When triturated with water, produces brown emulsion

119. Devil's Dung is synonym of ………..
    a. Colophoney
    b. Myrrh
    c. Asafoetida
    d. Cannabis

120. Which one of the following is biological source of asafoetida?
    a. Oleo – gum resin from rhizomes and roots of *Ferulafoetidu*
    b. Oleo – gum resin from rhizomes and roots of *Ferularubricaulis*
    c. Both a and b
    d. None

121. Which is the principle resin present in Asafoetida?
    a. Umbeliferone
    b. Asaresinotannol
    c. Umbellic acid
    d. None

122. Specific odour of asafoetida is due to presence of …………
    a. Phosphorous
    b. Sulphur
    c. Nitrogen
    d. Ammonia

123. Which resin acid is presented in colophony as chief constituent?
    a. Pimaric acid
    b. Abeitic acid
    c. Sapnic acid
    d. None

124. Industrially, colophony is used in preparation of
    a. Varnishes
    b. Soaps
    c. Paper sizing
    d. All of the above

125. Alcoholic solution of colophony is
    a. Basic to litmus
    b. Acidic to litmus
    c. Neutral to litmus
    d. None

Glycosides

126. Indian senna is…………
    a. Cassia anguistifolia
    b. Cassia acutifolia
    c. Both a and b
    d. None

127. Which type of glycosides are present in senna ?
    a. Anthracene glycosides
    b. Cyanogenetic glycosides
    c. cardiac glycosides
    d. None of the above

128. In sennoside B, the aglycone is …………
    a. Mesorotator
    b. Dextro rotatory
    c. Both a and b
    d. None

129. Which is the following naphthalene glycosides present in senna?

    a. Tinnivelley glycosides

    b. 6 – hydroxy musizin glycoside

    c. Both a and b

    d. None

130. Borntrager's test is employed for the presence of ....

    a. Cardiac glycosides

    b. Anthraquinones glycosides

    c. Saponin glycosides

    d. Coumarin glycosides

131. Which one of the following variety of senna is known as Alexandrian senna?

    a. Cassia anguistifolia

    b. Cassia acutifolia

    c. Cassia fistula

    d. Cassia tora

132. Aloin is mixture of

    a. C-10 glucosides

    b. C-11 glucosides

    c. C-10 glycosides

    d C- 11 glycosides

133. Modified anthraquinones test indicates presence of ....

    a. S – glycosides

    b. C- glycosides

    c. N- glycosides

    d. O – glycosides

134. Borax test is used for ....

    a. Identification of aloe

    b. Identification of senna

    c. Identification of Rhubarb

    d. Identification of Cascara

135. *Aloe babrbadensis* Miller is commonly known as ....

    a. Socotrine aloe

    b. Curacao aloe

    c. Cape aloe

    d. Zanzibar aloe

136. Socotrine aloe is prepared from ....

    a. Aloe ferox

    b. Aloe perryi

    c. Aloe barbadensis

    d. None of the above

137. Aloesin is ....

    a. Resin of aloe

    b. C – glucosyl chromome

    c. Both a and b

    d. None of the above

138. Amygdala amara is synonym of ....

    a. Prunus amygdalus

    b. Prunus communis

    c. Both and b

    d. None of the above

139. Biological source of bitter almond is ....

    a. Prunus amygdalus

    b. Prunus communis

    c. Both a and b

    d. None of the above

140. In bitter almond, bitter glycoside amygdalin is present in quantity ....
    a. 3 to 5 per cent
    b. 1 to 3 per cent
    c. 2 to 3 per cent
    d. 1 per cent

141. Amygdalin on hydrolysis yields ....
    a. Benzaldehyde
    b. Hydrocyanic acid
    c. Benzone
    d. Both a and b

142. Which hydrolysed component of amygdalin is very poisonous and unsuitable for internal consumption?
    a. Benzoin
    b. Hydrocyanic acid
    c. Benzaldehyde
    d. None of the above

143. Sweet almonds do not produce volatile oil because of ....
    a. Presence of amygdalin
    b. Absence of amygdalin
    c. Excessive drying process
    d. None

144. Bitter almonds are used as sedative due to content of ....
    a. Benzaldehyde
    b. Hydrocyanic acid
    c. Benzone
    d. Hydrochloric acid

145. Bitter almond oil is used in ....
    a. Demulcent skin lotion
    b. Hair shampoo
    c. Topical creams
    d. None

Iridoids, Other Terpenoids andNaphthaquinones

146. Which one of the following synonym of Artemisia?
    a. Devil's Dong
    b. Quinghao
    c. Santonica
    d. Sweet annie

147. The biological source of artemisia is ....
    a. Flower heads of *Artemisia cina*Berg
    b. Flower heads of *Artemisia brevifolia*Wall
    c. Flower heads of *Artemisia maritima* Linn
    d. All of the above

148. Artemisia belongs to family ....
    a. Compositae
    b. Asteraceae
    c. Leguminoseae
    d. Lilliaceae

149. The chief active constituent of Artemisia is ....
    a. Artemisin
    b. Santonin
    c. resin
    d. Santonic acid

150. Artemisia exerts strong ....
    a. Anti-malarial activity
    b. Antimicrobial activity
    c. Anthelmintic activity
    d. Anti inflammatory activity

151. Taxol is present in all parts of plants in quantity...........
    a. 0.7 – 1 %
    b. 0.007 – 0.1 %
    c. 0.07 – 0.01%
    d. 0.007 – 0.01%

152. Rare oxetane ring and amide side chain are present in ......
    a. Cephalomannine
    b. Baccatin
    c. Taxol
    d. None of the above

153. Taxol have been approved by USFDA for treatment of ....
    a. Leukemia
    b. Lung cancer
    c. Hodgkin's lymphoma
    d. Refractory ovarian cancer

154. Taxol exerts anticancer activity by ....
    a. Blocking cell cycle
    b. Enhancement of microtubule formation causing cell damage
    c. Prevents cell migration
    d. All of the above

155. In India, Taxus occurs in temperate region of Himalaya at an altitude ....
    a. 2000 – 3500 m
    b. 1000 – 2000 m
    c. 2000 – 3000 m
    d. 3500 – 4000 m

156. Tetraterpenoids are C_{40} compounds biosynthetically prepared by.......
    a. Tail to tail condensation of geranyl geraneol
    b. Head to tail condensation of geranyl geraneol
    c. Head to head condensation of geranyl geraneol
    d. None of the above

157. Which one of the medicinal plant is known as lipstick pod?
    a. Crocus sativus Linn
    b. Bixa orellana L
    c. Coleus forskohlii
    d. None of the above

158. Bixin protects against ....
    a. Infra red light
    b. Ultra violet light
    c. Radio waves
    d. Microwaves

159. Which bitter principle is present in saffron?
    a. Crocin
    b. Protocrocin
    c. Picrocrocin
    d. Crocetin

160. Biological source of gentian is ....
    a. Roots of Gentiana lutea belonging to family Genetinaceae
    b. Rhizomes of Gentiana lutea belonging to family Genetinaceae
    c. Both a and b
    d. None of the above

161. Gentian is used for.........
    a. Digestive disorders
    b. Skin diseases
    c. Respiratory diseases
    d. None

162. 'Gentian Brandy' comes from ....
    a. Fresh roots of plants
    b. Dried roots of plants
    c. fresh flowers of plants
    d. fresh fruits of plants

163. Which one of the following is active principle of Gentian?
    a. Genticin
    b. Gentiamarin
    c. Gentiopicrin
    d. Genitisic acid

164. Gentiopicrin is ....
    a. Iridoid
    b. Secoiridoid
    c. Sesquiterpene
    d. Monoterpene

165. On hydrolysis, Gentian yields ....
    a. Gentiogenin and glucose
    b. Mesogentiogenin and glucose
    c. Gentiopicrin and glucose
    d. Gentiogenin and sucrose

Answer Key

1. a	2. c	3. a	4. b	5. d	6. c	7. a	8. b	9. a	10. a
11. b	12. b	13. a	14. c	15. b	16. b	17. c	18. b	19. a	20. a
21. c	22. B	23. b	24. c	25. c					
26. b	27. b	28. a	29. d	30. d	31. b	32. c	33. b	34. a	35. a
36. b	37. b	38. c	39. c	40. b	41. c	42. d	43. a	44. b	45. a
26. b	27. a	28. b	29. c	30. c	31. b	32. b	33. d	34. b	35. a
36. c	37. a	38. c	39. d	40. a	41. c	42. b	43. a	44. b	45. d
66. b	67. c	68. b	69. a	70. a	71. d	72. b	73. a	74. a	75. b
76. c	77. b	78. b	79. d	80. c	81. c	82. a	83. b	84. a	85. A
86. a	87. c	88. a	89. b	90. c	91. b	92. c	93. a	94. a	95. b
96. d	97. a	98. c	99. b	100. b	101. c	102. a	103. d	104. d	105. b
106. d	107. c	108. c	109. a	110. c	111. c	112. b	113. c	114. a	115. c
116. a	117. a	118. a	119. c	120. c	121. b	122. b	123. b	124. d	125. b
126. d	127. c	128. c	129. a	130. c	131. c	132. b	133. c	134. a	135. c
136. a	137. a	138. a	139. c	140. c	141. b	142. b	143. b	144. d	145. b
146. c	147. d	148. a	149. b	150. c	151. d	152. c	153. d	154. d	155. a
156. a	157. b	158. b	159. c	160. c	161. a	162. a	163. c	164. a	165. b

Isolation, Identification and Analysis of Phytoconstituents

PCI Syllabus

Metabolic pathways in higher plants and their determination

- Brief study of basic metabolic pathways and formation of different secondary metabolites through these pathways- Shikimic acid pathway, Acetate pathways and Amino acid pathway.

- Study of utilization of radioactive isotopes in the investigation of Biogenetic studies

---

Chapter Content

3.1   Terpenoids

3.2   Glycosides

3.3   Alkaloids

3.4   Resins

3.1 Terpenoids

Artemisinin

Artemisinin is sesquiterpenoid lactone present in the flower heads of *Artemisia annua* of family *Asteraceae*. It is an antimalarial drug found to be effective against chloroquine resistant strains of *P. falciparum*.

IUPAC name: (3R,5aS,6R,8aS,9R,12S,12aR)-Octahydro-3,6,9-trimethyl-3,12-epoxy-12H-pyrano[4,3-j]-1,2-benzodioxepin-10(3H)-one

Extraction Methods

1. Take an accurately weighed quantity of *Artemisia* leaves powder. Extract the powder with petroleum ether.
2. Filter and evaporate the filtrate to dryness. Dissolve the residue in chloroform. Add acetonitrile to precipitate the impurities.
3. Separate the precipitate and discard. Concentrate the filtrate to dryness. Load this residue on a column chromatography by using chloroform-ethyl acetate mixture.
4. Monitor the elution by TLC. Combine the fraction containing artemisinin. Purify the crude artemisinin by using aqueous alcohol.

Identification Test

- Boil Artemisinin in 10ml of alcohol and filter. Add Sodium hydroxide to filtrate and heat. Red colour will indicate presence of Artemisinin.
- Dissolve 5 mg Artemisinin in about 0.5 ml of dehydrated ethanol, add about 0.5 ml of hydroxylamine hydrochloride and 0.25 ml of sodium hydroxide (~80 g/l). Heat the mixture in a water-bath to boiling, cool, add 2 drops of hydrochloric acid (~70 g/l) and 2 drops of ferric chloride (50 g/l); a deep violet colour is immediatelyproduced.
- Dissolve 5 mg Artemisinin in about 0.5 ml of dehydrated ethanol R, add 1.0 ml of potassium iodide (80 g/l), 2.5 ml of sulfuric acid (~100 g/l) and 4 drops of starch TS; a violet colour is immediately produced.
- Melting Point: 151–154 °C.

Thin Layer Chromatography: Stationary Phase: Silica Gel. G, Mobile Phase- toluene-ethyl acetate (95:5), Detection: PAB (0.25 g para-dimethylaminobenzaldehyde in 50 ml acetic acid and 5 ml of 10% phosphoric acid) reagent

Menthol

Menthol is a cyclic monoterpenoid alcohol can be obtained from number of mint plants belonging to family Lamiaceae. Mentha arvensis, M. aquatica, M. piperita, M.spicata are mostly use to isolate menthol.

Menthol's unique cooling sensation when inhaled, eaten, or applied to the skin is due to chemically triggered cold-sensitive TRPM8 receptors in the skin which is similar to capsaicin, the chemical responsible for the spiciness of hot chilis (which stimulates heat sensors, also without causing an actual change in temperature). Menthol's analgesic properties are mediated through a selective activation of κ-opioid receptors. Menthol is widely used in dental care as a topical antibacterial agent, effective against several types of streptococci and lactobacilli. It is preferred flavouring additives besides vanilla and citrus in many products due to cooling as well as anti-bacterial effects.

IUPAC Name (1R,2S,5R)-2-Isopropyl-5-methylcyclohexanol

Extraction Methods

1. Take the accurately weighed quantity of coarse powder of *Mentha piperita* parts just before flowering.
2. Extract the peppermint oil by water distillation method.
3. Separate the oil and allow cooling. Crystals of (-)menthol will separate out.
4. Collect the crystals by centrifugation.
5. Recrystallise menthol from acetone or any other low boiling point solvent.

Identification Test: Few drops of sample is mixed with 5ml of nitric acid and heated on water bath. Blue colour is developed within 5 minutes, after some time it becomes yellow which indicate the presence of menthol.

Properties:

- *Appearance*: White crystalline, solid at room temperature
- *Melting point:* 41 to 43°C
- *Odour* : Characteristic and pleasant

- ***Taste*** : Pungent followed by cooling sensation
- ***Solubility*** : Soluble in 70% alcohol, ether and chloroform, insoluble in water

Thin Layer Chromatography:*Stationary phase*: Silica gel G, Mobile phase : Chloroform, Detecting agent : Vanillin – sulphuric acid reagent and heat the plate at110^{0}C for 10 minutes, RF Value :0.48-0.62.

Citral

IUPAC Name : 3,7-dimethylocta-2,6-dienal

Citral is present in the oils of several plants, including lemon myrtle (90–98%), Litseacitrata (90%), Litsea cubeba (70–85%), lemongrass (65–85%), lemon tea-tree (70–80%), Ocimumgratissimum (66.5%), Lindera citriodora (about 65%), Calypranthesparriculata (about 62%), petitgrain (36%), lemon verbena (30–35%), lemon ironbark (26%), lemon balm (11%), lime (6–9%), lemon (2–5%), and orange.

Citral, or 3,7-dimethyl-2,6-octadienal or lemonal, is either a pair, or a mixture of terpenoids with the molecular formula C10H16O. The two compounds are geometric isomers. The E-isomer is known as geranial or citral A. The Z-isomer is known as neral or citral B.

Citral has been applied to food, cosmetics, and beverages as a natural ingredient for its passionate lemon aroma and flavor. Essential oils, which presenting citral have been demonstrated to show antimicrobial, antifungal, and antiparasitic characteristics, accomplishing citral a natural preservative.

Extraction Methods

Citral, a mixture of geranial (I) and neral (II) roughly in the ratio 5:3, is present in lemongrass oil generally to the extent of 70-80%. Citral has been separated from Iemongrass oil hy vacuum fractionation of the oil or by sodium bisulphite, and neutral sulphiteadductings methods. In vacuum fractionation the enrichment of citral happens and citral of 95% purity is generally obtained. Moreover removal of components like geraniol, nerol (IV) which have boiling points differing only by a few °C from that of citral is found to be difficult even when high efficiency fractionating columns are used. Being a mixture of α, β unsaturated aldehydes, citral is heat labile, and excessive heat treatment is likely to lead to rearrang-ements, polymeri-sation and eventual destruction of the material.

The other method of separation of citral, namely bisulphite adducting, also suffers from disadvantages. In this process, the lemongrass oil is shaken with a sodium bisulphite solution. The resulting crystalline solid is separated and purified by washing with alcohol or ether. The citral is regenerated by decomposition of the adduct with sodium carbonate, sodium hydroxide or hydrochloric acid. Even though the normal adduct is formed quantitatively and can be easily decomposed, quantitative regeneration of citral is difficult. Usually a loss of about 10-15% is encountered. The loss is reported to be due to the formation of a cyclic bisulphite compound in the presence of alkali from which the recovery of citral is found to be difficult.

IdentificationTest

Appearance: Clear pale yellow liquid

Odour: Strong lemon like odour

Taste: Lemon like taste

Solubility: Soluble in 3 parts of 70% alcohol, chloroform and fixed oil.

Insoluble in water.

Boiling point: 224-228O C

Chemical test: Alcoholic solution of Sudan red III and Tincture alkana is added to the sample. Red colour is appeared which indicate presence of citral.

Thin layer chromatography: Stationary phase: Silica gel G, Mobile phase: Chloroform, Detecting agent: 2, 4, dinitrophenyl hydrazine reagent, Color spots : Yellow to orange, RF Value : 0.51

3.2   Glycosides

Glycyrrhetinic Acid (Enoxolone)

Glycyrrhetinic acid or glycyrrhetic acid is a beta-amyrin pentacyclic triterpenoid aglycone obtained from the roots and stolones of *Glycyrrhiza glabra* (liquorice) ofthe family *Leguminoseae*. It mainly possesses antiulcer and expectorant (antitussive) properties as well as antiviral, antifungal, antiprotozoal, and antibacterial activities. Acetoxolone and Carbenoxolone derivatives of Glycyrrhetinic acid are also used for the treatment of peptic, esophageal and oral ulceration, andinflammation.

Glycyrrhizin = Glycyrrhetinicacid + 2 glucoronic acid

IUPAC name: (3β)-3-[(3-carboxypropanoyl)oxy]-11-oxoolean-12-en-30-oic acid

Extraction Methods

Method 1

1. Take the accurately weighed quantity of coarse powder of *Glycyrrhiza* roots.
2. Extract the powder with chloroform. Filter and discard the filtrate.
3. Extract the marc with 0.5 M H_2SO_4 for a few hours.
4. Filter and extract the filtrate with three portions of chloroform. Separate and combine the chloroform layers. Distill off the chloroform extract to yield a dry residue of glycyrrhetinic acid.

Method 2

1. Take the accurately weighed quantity of coarse powder of *Glycyrrhiza* roots.
2. Extract the powder with boiling water. Filter and concentrate to obtain a crude liquorice extract.
3. Dissolve this solid extract again in water and acidify with HCl to obtain a pH 3-3.4 to precipitate glycyrrhetinic acid.
4. Filter and wash the residue with water to yield glycyrrhetinic acid.

Method 3

1. Take 25 g of liquorice powder in a macerating flask; add 60 ml acetone and 3 ml dilute nitric acid.
2. Macerate for 3 hours and filter. Concentrate the filtrate under vacuum to yield a semisolid mass.
3. Add sufficient amount of dilute ammonia solution to precipitate ammonium glycyrrhizinate.
4. Filter the precipitate and wash with acetone. Again add ammonia to mother liquor and separate the remaining precipitate.
5. Combine and dry the precipitate to obtain ammonium glycyrrhizinate.

Identification Test

***Libermann-Burchard test*:** Extract solution mixed with few drops of acetic anhydride, boil and cool, concentrated sulphuric acid is added from the side of the test tube, A brown ring at the junction of two layers and the upper layer turns green indicates the presence of sterols and formation of deep red colour indicates the presence of triterpenoids.

***Salkowski's test*:** Dissolve the extract in chloroform with few drops of concentrated Sulfuric acid, shake well and allow to stand for some time, red colour appears in the lower layer indicates the presence of sterols and formation of yellow coloured lower layer indicating the presence of triterpenoids.

Rutin (Rutoside, Quercetin-3-O-Rutinoside and Sophorin)

Introduction

Rutin is a flavonoid obtained from various citrus fruits. It can be extracted from the leaves of *Fagopyrum esculentum,* also known as buckwheat, belonging to the family *Polygonaceae.* It is used as a capillary fragility factor as well as antibacterial and antioxidant properties.

Quercetin + rutinoside (Glucose + Rhamnose) = Rutin

IUPAC name: 2-(3,4-dihydroxyphenyl)-5,7-dihydroxy-3-[α-L-rhamnopyranosyl-(1→6)-β-D-glucopyranosyloxy]-4H-chromen-4-one

Extraction Methods

1. Take the accurately weighed quantity of a coarse powder of *fagopyrum* leaves.
2. Defat the powder with n-hexane. Extract the marc with 78% alcohol for 1 hour.
3. Filter and evaporate the solvent to obtain a dry residue.
4. Dissolve this residue in sufficient quantity of 30% acetone.
5. Filter and evaporate the filtrate to 1/4th of its original volume.
6. Add sufficient quantity of 5% aqueous solution of borax (until resulting pH is 7.5) with continuous stirring.

7. Add sufficient quantity of solid NaCl with stirring.

8. Filter and acidify the clear filtrate with phosphoric acid to bring pH to 5.5. Stir for 15 min. Filter and wash the residue with 20% NaCl solution.

9. Again filter and evaporate the filtrate to 500°C to 1/4th of its original volume.

10. Add HCl in hot condition to bring pH to 1.5. Cool and keep in refrigerator overnight.

11. Crystals of rutin will separate out. Collect and dry to calculate the percentage yield.

Identification Test

Shinoda test: Add magnesium powder and a few drops of concentrated HCl or H_2SO_4 to 2 ml of sample solution.

- Flavones, flavonols and xanthones: Orange, pink, red, and purple.

- Flavanones and flavononols: weak pink to magenta colors, or no color at all.

3.3 Alkaloids

Atropine

Atropine is a tropane alkaloid obtained from *Atropa belladonna* (deadly night shade), *Datura stramonium* (thorn apple), and *Hyoscyamus niger*(henbane) of family *Solanaceae*. It is found in many members of the *Solanaceae* family. It is a diastereomeric mixture of d-hyoscyamine and l-hyoscyamine with most of its physiological effects due to l-hyoscyamine. Its pharmacological effects are due to binding to muscarinic acetylcholine receptors. It is an antimuscarinic and anticholinergic agent.It is mydriatic and antidote in opium poisoning.Atropine is contraindicated in patients pre-disposed to narrow angle glaucoma.

IUPAC name: [(1R,5S)-8-methyl-8-azabicyclo[3.2.1]octan-3-yl] 3-hydroxy-2-phenylpropanoate

Extraction Methods

Method 1

1. Take the accurate weighed quantity of any of the above mentioned crude drug and powder to obtain coarse sized particles.

2. Moisten the powder with sodium carbonate solution. Extract the blended mixture in ether.

3. Filter the ether extract. Extract the filtrate with aqueous acetic acid.

4. Treat the aqueous acidic layer with solvent ether and separate both layers.

5. Discard ether layer. Treat acidic layer with sodium carbonate to obtain precipitates of tropane alkaloids.

6. Filter the precipitates and dry to obtain residue. Dissolve this residue in diethyl ether. Filter and concentrate the filtrate.

7. Atropine crystals will be separated out. Filter the crystals and dissolve in alcohol containing NaOH solution so that hyoscyamine can be converted to atropine.

8. Recrystallize the atropine sulphate from acetone.

Method 2

1. Take the accurate weighed quantity of belladonna leaves and powder to obtain coarse sized particles.

2. Extract the powder with 95% ethanol. Filter and distill off solvent to syrupy mass.

3. Add sufficient quantity of 1% HCl to syrupy mass. This will remove resinous matter.

4. Now, add petroleum ether. Shake well and made alkaline with ammonia.

5. Extract this mixture with chloroform. Separate chloroform layer and add dilute acid. Shake and separate chloroform layer. Add ammonia to make solution alkaline.

6. Evaporate chloroform layer to $1/4^{th}$ of its original volume. Add oxalic acid which will separate out crystals of atropine and hyoscyamine oxalate.

7. To separate hyoscyamine from atropine, dissolve crystals in acetone or ether. Then, filter it. Filtrate will have hyoscyamine oxalate crystals and residue will be of atropine oxalate.

Method 3

1. Take the accurate weighed quantity of powder of belladonna roots or datura leaves.

2. Extract the powder with 95% ethanol. Filter it and add potassium carbonate to filtrate to convert hyoscyamine to atropine.

3. Now, add chloroform to this solution. Shake well and separate chloroform layer.

4. Evaporate chloroform to dryness. Dissolve residue in dilute H_2SO_4.

5. Add potassium carbonate to precipitate atropine. Separate precipitate and extract with ether. Add oxalic acid to ether solution to give atropine oxalate crystals.

Indentification Test

Atropine	**Vitali-Morin's test** Take a small quantity of solid atropine and add 2 drops of concentrated nitric acid and mix it in an evaporating dish and evaporate to dryness on water bath to obtain faintly yellow residue. Cool well. Then, dissolve the residue in 1 ml of acetone. Add few drops of freshly prepared solution of alcoholic potassium hydroxide.	A bright purple color is formed which changes to red color and fades gradually
	Gerrard reaction Take 1 ml of 2% solution of mercuric chloride in 50% alcohol and add few mg of atropine.	A red color will be produced.

Caffeine

Most of the people start their morning with the cup of tea which contains the alkaloid caffeine. Caffeine is a naturally occurring substance found in the leaves, seeds or fruits of over 63 plant species worldwide, and is part of a group of compounds known as methylxanthines. The most commonly known sources of caffeine are coffee, cocoa beans, cola nuts, and tea leaves. Caffeine is a bitter, white crystalline xanthine alkaloid that is a psychoactive stimulant drug. Caffeine is pseudo alkaloid as it is not biosynthesized from amino acids, but gives all alkaloid identification tests positive. Extraction of caffeine is a simple decoction process where caffeine is first dissolved in water and separated by chloroform. Tea leaves contain 1%-4% of caffeine while coffee seeds contain 1%-2% of caffeine.

IUPAC name: 1,3,7-trimethylpurine-2,6-dione

Extraction Methods

Method 1

1. Weigh 10 g of tea leaves and transfer to 250 ml distilled water.

2. Boil the water for 30 minutes with occasional stirring.

3. Then allow tocool and filter the solution.

4. Take filtrate in a separating funnel and to it, add 100 ml chloroform. Shake vigorously so that total caffeine will be transferred to chloroform.

5. Separate chloroform layer. Evaporate chloroform over water bath. White caffeine crystals will collect at the bottom.

Method 2

1. Powder the tea leaves to coarse sized particles and extract with water for half an hour.

2. Filter the extract. Treat with lead acetate to precipitate tannins. Filter the solution and add dilute sulphuric acid to the solution to remove the excess lead acetate in the form of lead sulphate.

3. Filter and add charcoal to remove the coloring matter. Filter this solution.

4. Take this filtrate in a separating funnel and extract with 3 portions of chloroform with mild shaking. Separate and combine the chloroform layers and evaporate to precipitate white crystals of caffeine. Recrystallize with alcohol.

Method 3

1. Powder the tea leaves to coarse sized particles and extract with 5% aqueous sodium carbonate solution in a beaker for 30 min.
2. Filter the above extract and neutralize the filtrate with 10% H_2SO_4 solution.
3. Again filter the solution with suitable filter aid. Wash the marc with dichloromethane.
4. Take the above mixture in a separating funnel to separate dichloromethane layer.
5. Evaporate dichloromethane layer to obtain caffeine crystals.

Method 4

1. Powder the tea leaves to coarse sized particles.
2. Heat the powder in 100 ml beaker which should be covered with funnel having closed end.
3. White caffeine crystals are deposited at the end of the glass funnel.
4. Care must be taken to avoid the charring of crystals.
5. Collect the crystals; weigh and calculate the percentage yield.

Identification Test

Caffeine:

Murexide test

Treat caffeine powder with HCl and KCl, heat to dryness, and expose residue to dilute ammonia.Purple color

Tannic acid test

Treat caffeine with tannic acid.White precipitate

Quinine and Quinidine

(−) quinine, (+) quinidine, cinchonine and cinchonidine are the popular quinoline alkaloids obtained from the dried bark of *Cinchona officinalis*, *C. calisaya*, *C. ledgeriana*, and *C. succirubra*offamily *Rubiaceae*. All these constituents have antimalarial properties. Quinidine is class I antiarrhythmic agent.

Quinine   Quinidine

Quinine IUPAC name:(R)-(6-Methoxyquinolin-4-yl)((2S,4S,8R)-8-vinylquinuclidin-2-yl) methanol

Quinidin IUPAC name: (S)-(6-Methoxyquinolin-4-yl)((2R,4S,8R)-8-vinylquinuclidin-2-yl)methanol

Extraction Methods

Method 1

1. Powder the cinchona bark to coarse sized particles and weigh accurately.
2. Moisten cinchona powder with alcoholic calcium hydroxide or calcium oxide or potassium hydroxide (20%) and sodium hydroxide solution (5%).
3. Keep this mixture in this form for few hours so that alkali can convert cinchona alkaloids to free bases.
4. Being non-polar in nature, these free bases can now be extracted with benzene for 6 hours in soxhlet extractor.
5. After complete extraction, filter the benzene extract and add 5% sulphuric acid.
6. Shake the above mixture in a separating funnel. Separate the aqueous layer from the benzene layer. Discard benzene layer. (In the above step, the free bases of cinchona alkaloid reacts with H_2SO_4 and forms water soluble sulphate salts).
7. Adjust the pH of aqueous layer to 6.5 with sodium hydroxide. Precipitates of quinine sulphate are formed, filter and recrystallize with hot water. Weigh and determine percentage yield and melting point.
8. Dissolve this quinine sulphate in dilute sulphuric acid. Add ammonia to crystallize quinine as sharp needles. Separate these crystals from mother liquor.
9. Treat this mother liquor with ammonia to precipitate quinidine, cinchonine and cinchonidine. Separate the precipitate and extract with ether.
10. The ether insoluble portion consists of cinchonine, while the ether insoluble portion is a mixture of quinidine and cinchonidine.
11. To separate quinidine and cinchonidine, add dilute HCl and then sodium potassium tartrate (25%) to ether soluble part. Filter the solution to separate cinchonidine tartrate precipitate which can be recrystallize from hot alcohol.
12. To the above filtrate, add potassium iodide solution to form precipitation of quinidine hydroiodide which is also recrystallized from alcohol.

Method 2

1. Powder the cinchona bark to coarse sized particles and weigh accurately.
2. Moisten cinchona powder with alcoholic KOH (20%). Keep this mixture aside to dry.
3. Extract this mass with benzene for 6 hours in soxhlet extractor.
4. After complete extraction, filter the benzene extract and shake with 5% sulphuric acid.

5. Filter the solution using filter paper coated with charcoal.

6. Basify to pH 8.5 by using ammonia solution. Extract the solution with 3 portions of chloroform.

7. Combine and evaporate the chloroform layer to obtain dry residue of total cinchona alkaloids.

8. Dissolve this residue in hot water containing small quantity of charcoal. White quinine sulphate crystals are separated.

Indentification Test

Cinchona	**Erythroquinine test** Take dilute acidic solution of quinine; add few drops of bromine water and a drop of 10% solution of potassium ferrocyanide. Then, add a drop of strong ammonia solution.	Red color which is taken up by chloroform layer after addition of 1 ml of chloroform.
	Thalloquin test Take powder and add one drop of dilute sulphuric acid and 1 ml of water. Then, add bromine water to get yellow color and finally add dilute NH_3 solution.	Emerald green color
	Vapor test Take powder and add glacial acetic acid to it. Now, heat it.	Purple vapors are produced
Quinine and quinidine	Take a little quantity of alkaloids and 0.25 ml of bromine water. Mix well and add 1 ml of water and 1 ml of chloroform. Allow it to stand for a few minutes. Then, add one drop of 10% potassium ferrocyanide and shake well. Add 3 ml of NaOH solution and shake well.	• Quinidine: Chloroform layer becomes red • Quinine: Chloroform layer remains colorless

Reserpine (Raudixin, Serpalan, Serpasil)

Reserpine is an indole alkaloid obtained from the roots of *Rauvolfia serpentina*, *R. micmntha*, *R. vomi-form* and *R. tetraphylla* of family *Apocynaceae*. Reserpine is used as potent hypotensive and sedative agent. Reserpine irreversibly blocks the vesicular monoamine transporter (VMAT) and causes depletion of intracellular norepinephrine, serotonin, and dopamine.It may take days to weeks to replenish the depleted VMAT, and thus, reserpine's effects are long-lasting. Depletion of dopamine can lead to drug-induced parkinsonism too.

IUPAC name: methyl (3β,16β,17α,18β,20α)-11,17-dimethoxy-18-[(3,4,5-trimethoxybenzoyl)oxy]yohimban-16-carboxylate

Extraction Methods

1. Take the accurately weighed quantity of dried rauvolfia roots and powder to obtain coarse sized particles.
2. Extract the powder with 90% alcohol in soxhlet apparatus.
3. Filter the alcoholic extract and evaporate to dryness.
4. Extract the residue with the mixture of ether: chloroform: 90% alcohol (20:8:2.5).
5. Filter and add dilute ammonia to filtrate.
6. Add water to precipitate the crude alkaloid mixture. Collect the precipitate and dry.
7. Dissolve the residue in 0.5N H_2SO_4. Add ammonia to the above solution to liberate alkaloid salts.
8. Extract the solution with 3 portions of chloroform. Evaporate the chloroform to obtain total rauvolfia alkaloids.
9. Separate the reserpine from the crude extract by column chromatography separation technique.

Identification Test

Reserpine	Treat the reserpine powder with vanillin-acetic acid solution.	Violet red color

3.4 Resins

Podophyllotoxin

Podophyllotoxin is the lactone resin present in the root and rhizome of *Podophyllum hexandrum* belonging to the family *Berberidaceae*. It is used as an antiproliferative agent. Synthetic derivatives include etoposide, teniposide, and etopophos.

Podophyllotoxin 2

IUPAC name: (10R,11R,15R,16R)-16-hydroxy-10-(3,4,5-trimethoxyphenyl)-4,6,13-
trioxatetracyclo[7.7.0.03,7.011,15]hexadeca-1,3(7),8-trien-12-one

Extraction Methods

1. Take accurately weighed quantity of rhizome/roots of *Paeonia emodi*with methanol. Filter and evaporate to semisolid mass.

2. Dissolve semisolid mass into acidic water. Precipitate is formed which should be allowed to settle for at least 2 hours.

3. Filter and wash the filtrate with cold water. Collect the residue, wash with acidified water, and dry to obtain dark brown amorphous powder.

4. Extract the residue with hot alcohol. Filter and evaporate to dryness. Recrystallize the residue in benzene to yield podophyllotoxin.

Identification Test: Treat podophyllotoxin with 50% Sulphuric acid. It will show violet-blue colour

Thin Layer Chromatography: Stationary phase: Silica gel G, Mobile phase: Chloroform: Methanol (90:10) for about 6cm (Only glycosides are separated but aglycone like podophyllotoxin remains in the region of the front. The same plate is again eluted with more weakly polar Solvent Chloroform: Acetone (65:35) upto 12cm. Detecting agent: Spray with methanol Sulphuric acid and heat 10 minutes at 110^0 C. RF Value: 0.65 Yellow spot

Curcumin

Introduction

Curcumin is diarylheptanoid and principal curcuminoid present in the rhizomes of *Curcuma longa* offamily Zingiberaceae commonly called as turmeric. This is used as an anti inflammatory, antiseptic, anticancer and hepatoprotective agent. It is also a common food colourant. Curcumin is a bright-yellow color and may be used as a food coloring. As a food additive,its E number is E100.

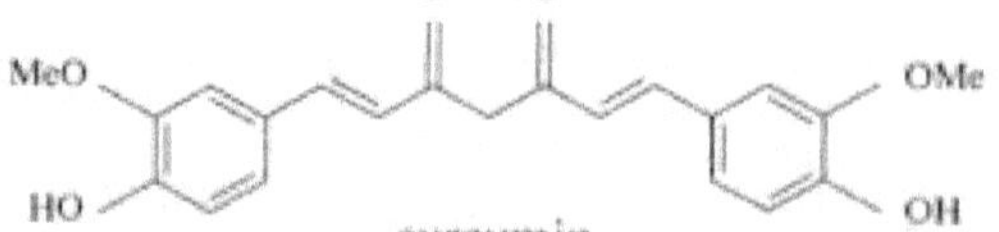

curcumin

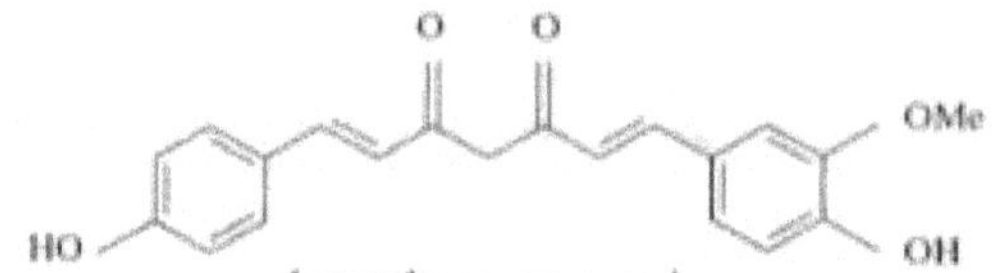

demethoxycurcumin

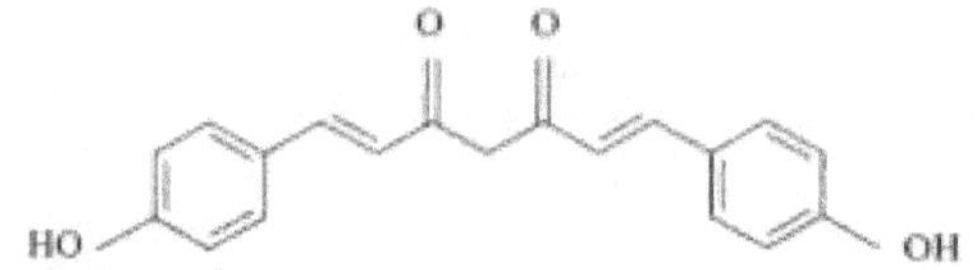

bisdemethoxycurcumin

Curcumin IUPAC name: (1E,6E)-1,7-Bis(4-hydroxy-3-methoxyphenyl)-1,6-heptadiene-3,5-dione

Extraction Methods

Method 1

1. Take accurately weighed quantity of turmeric powder.
2. Extract with n-hexane for 2 hrs.
3. Discard the n-hexane extract and extract marc with acetone for 2 hrs.
4. Distill off the acetone and dry the crystals of curcumin.
5. Recrystallise the curcumin from hot ethanol.

Method 2

1. Take accurately weighed quantity of turmeric powder.
2. Extract with n-hexane for 2 hrs.
3. Discard the n-hexane extract and extract marc with methanol for 2 hrs.
4. Distill off the methanol to obtain solid residue.
5. Dissolve this residue in any suitable alkali, filter and acidify with HCl to precipitate crystals of curcumin.

Identification Test: Treat sample with acetic anhydride and conc. H2SO4, it gives violet colour. Then observe under UV light, red fluorescence confirms presence of curcuminoids.

Subjective Questions

1. How to isolate volatile oils and separate various constituents from crude oil?

2. How to isolate, identify and analyse Atropine or Curcumin?

3. How to isolate, identify and analyse Artemisinin or Glycyrrhizin?

4. How to isolate and identify

    a. Rutin   b. Quinine   c.Reserpine   d.   Caffeine

Multiple Choice Questions (MCQs)

1. L- Menthol is commercially isolated by which method?
    a. Fractional distillation
    b. Steam distillation
    c. Fractional liberation
    d. Crystallization

2. In litmus paper test, litmus turns ....when comes in contact with menthol
    a. Acidic
    b. Basic
    c. Neutral
    d. None

3. During analysis of menthol, at what temperatures optical rotation should be performed to get accurate result?
    a. 30°C
    b. 27°C
    c. 25°C
    d. 20°C

4. Specific gravity of menthol is ....
    a. 0.9 – 0.912g
    b. 1g
    c. 1.42g
    d. 1.2g

5. For commercial scale isolation, which apparatus is used generally?
    a. Galvanised iron still
    b. Mild steel still
    c. Both a and b
    d. None of the above

6. What factors should be avoided while preparing mentha oil?
    a. Drying in sunlight
    b. Fermentation
    c. Both a and b
    d. None

7. Standard refractive index value for menthol is ....
    a. 1.4590 – 1.4650
    b. 1.5321 – 1.5430
    c. 1.3590 – 1.3650
    d. 1.2590 – 1.2650

8. The thick section of G. glabra drug or powder shows deep yellow colour on addition of
    a. 80% hydrochloric acid
    b. 80% sulphuric acid
    c. 80% nitric acid
    d. 80% dil. Sulphuric acid

9. The standard ash value for peeled drug of *Glycyrrhiza glabra* is ....
   a. Not more than 10 per cent
   b. Not less than 10 per cent
   c. Not more than 6 per cent
   d. Not less than 6 per cent

10. Standard acid insoluble ash value for peeled liquorice drug is ....
    a. Not more than 2 per cent
    b. Not more than 1 per cent
    c. Less than 2 per cent
    d. Less than 2 per cent

11. Citral is isolated from ....
    a. Cymbopogon citratis
    b. Mentha piperata
    c. Carm carvi
    d. Elettaria cardamomum

12. Citral is isolated by ....
    a. Fractional liberation
    b. Fractional distillation
    c. steam distillation
    d. Crystallization

13. Standard optical rotation value for lemongrass oil is ....
    a. 3° to 1°
    b. -3° to +1°
    c. 4° to +1°
    d. -4° to +1°

14. Standard specific gravity value of lemongrass oil is ....
    a. 0.990 -1.040
    b. 0.892 – 0.909
    c. 1.042 – 1.102
    d. 1.132 – 1.145

15. Standard refractive index value of Lemongrass oil is ....
    a. 1.4808 – 1.4868
    b. 1.4508 – 1.4568
    c. 1.4008 – 1.4068
    d. 1.4308 – 1.4368

16. Artemisin is isolated from ....
    a. Flower heads of *Artemisia cina* Berg
    b. Whole plant of *Artemisia annua* Linn.
    c. Both a and b
    d. None of the above

17. When alcoholic extract of Artimisia is treated with sodium hydroxide, the liquid develops
    a. Yellow colour
    b. Green colour
    c. Red colour
    d. Brown colour

18. Menthol when treated with nitric acid solution, develops
    a. Red colour
    b. Blue colour
    c. yellow colour
    d. green colour

19. In India, Menthol is manufactured by ....
    a. S. H. Kelkar
    b. Procter and Gamble Ltd
    c. Bhavana Chemicals
    d. All of these

20. Menthol is ....
    a. Amorphous in nature
    b. Crystalline in nature
    c. Liquid in nature
    d. None

21. Which one of the following is identification test for Atropine?

    a. Baljet test
    b. Vitali Morin test
    c. Killer kiliani test
    d. salkowaski test

22. Which one of the following is synthetic derivative of atropine used commercially?

    a. Hyoscine
    b. Hyoscyamine
    c. Homotropine
    d. None of the above

23. Standard total ash value for *Atropa belladonna* is

    a. 15 per cent
    b. 13 per cent
    c. 14 per cent
    d. 15 per cent

24. At the time of isolation, belladonna powder is first moistened with ....

    a. Na2CO3
    b. NaHCO3
    c. Nacl
    d. KOH

25. For crystalization of alkaloid atropine, which solvent is used?

    a. Acetone
    b. Ether
    c. Both a and b
    d. None

26. For extraction of quinine, which solvent is used preferably?

    a. Ethanol
    b. Methanol
    c. Toluene
    d. Diethy ether

27. Quinine is separates from extract in the form of ....when treated with dil. Sulphuric acid

    a. Qunine phosphate
    b. Qunine sulphate
    c. Quinine hydrate
    d. None

28. The test in which quinine gives emerald green colour with bromine water and dilute ammonia solution is known as

    a. Vitali Morine test
    b. Killer killiani test
    c. Borntrager test
    d. Thalleoquin test

29. Solubility of quinine in water is

    a. 1 part in 1000 parts of water
    b. 1 part in 900 parts of water
    c. 1 part in 810 parts of water
    d. 1 part in 700 parts of water

30. In UV light, quinine shows

    a. Red fluorescence
    b. Blue fluorescence
    c. Green fluorescence
    d. None

31. In the seeds, caffeine is present in the form of

    a. Salt of caffeotannic acid
    b. Salt of chlorogenic acid
    c. Salt of valerianic acid
    d. None

32. For synthesis of caffeine, which one of the following is used as starting material?
    a. Uric acid
    b. Tryptophan
    c. Aniline
    d. Tryptophan

33. Which one of the following method is used for commercial method for production of caffeine?
    a. Sublimation
    b. Steam distillation
    c. Fractional crystallization
    d. Fractional crystallization

34. Which one of the following test is used for identification of caffeine?
    a. Thalleoquine test
    b. Murexide test
    c. Vitali Morin test
    d. Wagner's test

35. Reserpine is extracted from *Rauwolfia serpentina* by which method ?
    a. Soxhlet extraction
    b. Maceration
    c. Steam distillation
    d. None of the above

36. T.S of *Rauwolfia serpentina* along medullary ray develops................colour when treated with nitric acid
    a. Blue
    b. Yellow
    c. Violet
    d. Red

37. Pwdered rauwolfia when treated with sulphuric acid and p – dimehyl amino benzaldehyde Develops
    a. Violet to red colour
    b. Red to violet colour
    c. violet to green colour
    d. green to violet colour

38. Which solvent is widely used for extraction of reserpine?
    a. Ethanol
    b. Chloroform
    c. Methanol
    d. diethyl ether

39. Which one of the microscopic character is ABSENT in Rauwolfia serpentina?
    a. Calcium oxalate crystals
    b. Stone cells
    c. Phloem fibers
    d. both b and c

40. Reserpine is ester derivative of
    a. Methyl reserpate
    b. Trimethoxybenzoic acid
    c. Trimethoxycinnamic acid
    d. Both a and c

41. Podophyllum resin contains podophyllotoxin
    a. 40 – 50 %
    b. 30 – 40 %
    c. 10 – 20 %
    d. 20 – 30%

42. Solvent used for extraction of podophyllotoxin is
    a. Ethanol
    b. Benzene
    c. Acetone
    d. Chloroform

43. Podophyllotoxin is extracted by which method?
    a. Soxhlet extraction
    b. Percolation
    c. Decoction
    d. Infusion

44. Which solvent is used for isolation of resin podophyllotoxin from extractby precipitation?
    a. Ethanol
    b. Acetone
    c. Diethyl ether
    d. water

45. Podophyllotoxin exerts anti mitotic activity due to presence of ………………….in trans position in structure
    a. Hydroxyl and lactone group
    b. Phenolic and lactone group
    c. Acetyl and lactone group
    d. None

46. When hydroxyl and lactone groups are in trans position in structure, podophyllotoxin exerts
    a. Anti inflammatory activity
    b. Anti diabetic activity
    c. Purgative activity
    d. antimitotic activity

47. When alcoholic solution of Indian podophyllotoxin reacts with potassium hydroxide solution, gives
    a. No Precipitate
    b. Stiff gely
    c. brown clour
    d. Red colour

48. When alcoholic solution of American podophyllotoxin reacts with potassium hydroxide solution, gives
    a. No Precipitate
    b. Stiff gely
    c. brown clour
    d. Red colour

49. Podophyllotoxin is NOT soluble in
    a. Cold water
    b. Chloroform
    c. alcohol
    d. ether

50. Podophyllotoxin is partially soluble in
    a. Cold water
    b. Hot water
    c. Chloroform
    d. Both b and c

51. Which solvent is highly suitable for extraction of curcumin?
    a. Acetone
    b. Water
    c. Ethyl acetate
    d. Methanol

52. Which method is highly suitable for extraction of curcumin from Curcuma longa ?
    a. Continous hot percolation
    b. Cold percolation
    c. Maceration
    d. Infusion

53. When powdered curcuma longa is treated with sulphuric acid, it gives
    a. Brown colour
    b. Crimson colour
    c. Yellow colour
    d. red colour

54. The aqueous solution of turmeric when treated with boric acid, it gives
    ....which on addition of alkali changes to ....
    a. Greenish blue, reddish brown
    b. Reddish brown, Greenish blue
    c. violet, fluorescent blue
    d. Reddish brown, Violet

55. Standard water soluble extractive value of curcuma is
    a. Not less than 9 per cent
    b. Not less than 10 per cent
    c. Not more than 9 percent
    d. Not more than 10 percent

56. Why recrystallization of curcumin is a difficult step in extraction and isolation process?
    a. Because of presence of volatile oil/oleo resin in drug
    b. Because of Lipophillic nature of curcumin
    c. Because curcumin gets dissolved in volatile during recrystallization process
    d. None of the above

57. Which spraying agent is used for detection of isolated curcumin spots on TLC plates ?
    a. Vanillin sulphuric acid
    b. Anisaldehyde sulphuric acid
    c. Methanolic sulphuric acid
    d. None

58. Melting point of pure curcumin is
    a. 183°C
    b. 283°C
    c. 200°C
    d. 171 °C

59. Curcuminoids isolated from ethyl acetate extract of turmeric have shown modest
    a. Antimicrobial activity
    b. HIV -1&2 protease inhibitory activity
    c. Antihelmithic activity
    d. Anticancer activity

Answer Key

1. b	2. c	3. c	4. a	5. c	6. c	7. a	8. b	9. c	10. a
11. a	12. c	13. b	14. b	15. a	16. a	17. c	18. b	19. d	20. b
21. b	22. c	23. c	24. a	25. c	26. c	27. b	28. d	29. c	30. b
31. a	32. a	33. a	34. b	35. a	36. d	37. a	38. b	39. d	40. d
41.a	42.a	43.b	44.d	45.a	46.d	47.b	48.a	49.a	50. d
51.a	52.a	53.b	54.b	55.b	56.c	57.a	58.a	59.b	

Unit 4

Industrial Production, Estimationand Utilization

PCI Syllabus

- Industrial production, estimation and utilization of the following phytoconstituents

Chapter Content

4.1   Forskolin

4.2   Sennoside

4.3   Artemisinin

4.4   Diosgenin

4.5   Digoxin

4.6   Atropine

4.7   Podophyllotoxin

4.8   Caffeine

4.9   Taxol

4.10   Vincristine and Vinblastine

Introduction

Several analytical techniques from chromatography like TLC, HPTLC, HPLC, GLC, and Spectroscopy like UV-Visible, colorimetry are commonly useful for the quantitative determination of phytochemicals in plants, extracts and derived formulations. This chapter is focusing on industrial production, estimation and utilization of the following phytoconstituents.

Menthol	**Utilization**: Menthol is a cyclic monoterpenoid alcohol can be obtained from number of mint plants belonging to family Lamiaceae. Mentha arvensis, M. aquatica, M. piperita, M.spicata are mostly use to isolate menthol. Menthol's unique cooling sensation when inhaled, eaten, or applied to the skin is due to chemically triggered cold-sensitive TRPM8 receptors in the skin which is similar to capsaicin, the chemical responsible for the spiciness of hot chilis (which stimulates heat sensors, also without causing an actual change in temperature). Menthol's analgesic properties are mediated through a selective activation of κ-opioid receptors. Menthol is widely used in dental care as a topical antibacterial agent, effective against several types of streptococci and lactobacilli. It is preferred flavouring additives besides vanilla and citrus in many products due to cooling as well as anti-bacterial effects.

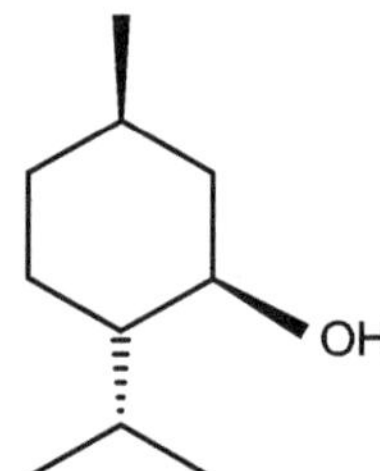

IUPAC Name (1*R*,2*S*,5*R*)-2-Isopropyl-5-methylcyclohexanol

Isolation: Take the accurately weighed quantity of coarse powder of *Mentha piperita* parts just before flowering. Extract the peppermint oil by water distillation method. Separate the oil and allow cooling. Crystals of (-)menthol will separate out. Collect the crystals by centrifugation. Recrystallise menthol from acetone or any other low boiling point solvent

Identification: Few drops of sample is mixed with 5ml of nitric acid and heated on water bath. Blue colour is developed within 5 minutes, after some time it becomes yellow which indicate the presence of menthol.

Properties:
- *Appearance*: White crystalline, solid at room temperature
- *Melting point:* 41 to 43°C

Contd...

	• *Odour* : Characteristic and pleasant • *Taste* : Pungent followed by cooling sensation • *Solubility* : Soluble in 70% alcohol, ether and chloroform, insoluble in water **Thin Layer Chromatography**: Stationary phase : Silica gel G, Mobile phase: Chloroform, Detecting agent : Vanillin – sulphuric acid reagent and heat the plate at110^{0}C for 10 minutes, RF Value : 0.48-0.62
Glycyrhetinic acid	**Utilization:**Glycyrrhetinic acid or glycyrrhetic acid is a beta-amyrin pentacyclic triterpenoid aglycone obtained from the roots and stolones of *Glycyrrhiza glabra (*liquorice) ofthe family *Leguminoseae*. It mainly possesses antiulcer and expectorant (antitussive) properties as well as antiviral, antifungal, antiprotozoal, and antibacterial activities. Acetoxolone and Carbenoxolone derivatives of Glycyrrhetinic acid are also used for the treatment of peptic, esophageal and oral ulceration, and inflammation. Glycyrrhizin = Glycyrrhetinic acid + 2 glucoronic acid IUPACname:(3β)-3-[(3-carboxypropanoyl)oxy]-11-oxoolean-12-en-30-oic acid **Isolation**: Take the accurately weighed quantity of coarse powder of *Glycyrrhiza* roots. Extract the powder with chloroform. Filter and discard the filtrate. Extract the marc with 0.5 M H_2SO_4 for a few hours. Filter and extract the filtrate with three portions of chloroform. Separate and combine the chloroform layers. Distill off the chloroform extract to yield a dry residue of glycyrrhetinic acid. **Identification:** ***Libermann-Burchard test:*** Extractsolution mixed with few drops of acetic anhydride, boil and cool, concentrated sulphuric acid is added from the side of the test tube, A brown ring at the junction of two layers and the upper layer turns green indicates the presence of sterols and formation of deep red colour indicates the presence of triterpenoids. ***Salkowski's test***: Dissolve the extract in chloroform with few drops of concentrated Sulfuric acid, shake well and allow to stand for some time, red colour appears in the lower layer indicates the presence of sterols and formation of yellow coloured lower layer indicating the presence of triterpenoids.

Rutin	**Utilization:**Rutin is a flavonoid obtained from various citrus fruits. It can be extracted from the leaves of *Fagopyrum esculentum,* also known as buckwheat, belonging to the family *Polygonaceae.* It is used as a capillary fragility factor as well as antibacterial and antioxidant properties. Quercetin + rutinoside (Glucose + Rhamnose) = Rutin **IUPAC** name:2-(3,4-dihydroxyphenyl)-5,7-dihydroxy-3-[α-L-rhamnopyranosyl-(1→6)-β-D-glucopyranosyloxy]-4H-chromen-4-one **Isolation:** Take the accurately weighed quantity of a coarse powder of *fagopyrum* leaves. Defat the powder with n-hexane. Extract the marc with 78% alcohol for 1 hour. Filter and evaporate the solvent to obtain a dry residue. Dissolve this residue in sufficient quantity of 30% acetone. Filter and evaporate the filtrate to 1/4th of its original volume. Add sufficient quantity of 5% aqueous solution of borax (until resulting pH is 7.5) with continuous stirring. Add sufficient quantity of solid NaCl with stirring. Filter and acidify the clear filtrate with phosphoric acid to bring pH to 5.5. Stir for 15 min. Filter and wash the residue with 20% NaCl solution. Again filter and evaporate the filtrate to 500°C to 1/4th of its original volume. Add HCl in hot condition to bring pH to 1.5. Cool and keep in refrigerator overnight. Crystals of rutin will separate out. Collect and dry to calculate the percentage yield. **Identification:**ShinodaTests to the extract solution, add few fragments of magnesium ribbon and concentrated HCl drop wise, yellowish; yellow-orange occasionally orange colour appears after few minutes
Quinine	**Utilization:**(−) quinine,(+) quinidine, cinchonine and cinchonidine are the popular quinoline alkaloids obtained from the dried bark of *Cinchona officinalis, C.calisaya, C.ledgeriana,* and *C.succirubra*offamily *Rubiaceae.* All these constituents have antimalarial properties. Quinidine is classI antiarrhythmic agent.

Contd...

Quinine IUPAC name:(R)-(6-Methoxyquinolin-4-yl)((2S,4S,8R)-8-vinylquinuclidin-2-yl) methanol

Quinidin IUPAC name: (S)-(6-Methoxyquinolin-4-yl)((2R,4S,8R)-8-vinylquinuclidin-2-yl)methanol

Isolation: Powder the cinchona bark to coarse sized particles and weigh accurately. Moisten cinchona powder with alcoholic calcium hydroxide or calcium oxide or potassium hydroxide (20%) and sodium hydroxide solution (5%). Keep this mixture in this form for few hours so that alkali can convert cinchona alkaloids to free bases. Being non-polar in nature, these free bases can now be extracted with benzene for 6 hours in soxhlet extractor. After complete extraction, filter the benzene extract and add 5% sulphuric acid. Shake the above mixture in a separating funnel. Separate the aqueous layer from the benzene layer. Discard benzene layer. (In the above step, the free basesof cinchona alkaloid reacts with H_2SO_4 and forms water soluble sulphate salts). Adjust the pH of aqueous layer to 6.5 with sodium hydroxide. Precipitates of quinine sulphate are formed, filter and recrystallize with hot water. Weigh and determine percentage yield and melting point.

Identification: Thelleoquin test: Take powder and add one drop of dilute sulphuric acid and 1 ml of water. Then, add bromine or chlorine water or calcium hypchlorite/ bleaching powder (modified thelleoquin test) to get yellow color and finally add dilute NH3 solution to get Emerald green color.

Thin Layer Chromatography (TLC): Stationary phase: Silica gel G, Mobile phase: Chloroform: Diethyl amine (9:1), Detecting agent: Dragendorff's reagent, RF Value: 0.17

High Pressure Liquid chromatography (HPLC): Method: Isocratic, Stationary phase: C18 coloum, Mobile Phase: Methanol: Acetonitrle-0.1mol/L: ammonia: acetone (45:15:40), Detection: Fluorescence at excitation 325nm, Emission : 375nm

Estimation Method:

Method 1: Bromocresol green Reagent method

Method 2: Flurometric analysis methods

Reagents:

➢ Quinine Sulfate, Stock Standard: 100 mg/L. Weight out 30.2 mg quinine sulfate.2H2O and dissolve in 250 mL 0.1 N H2SO4.

Contd...

> Quinine Standards, Working: 5, 10 and 15 mg/L. Using 100 mL volumetric flask, dilute appropriate volumes of the quinine stock standard using 0.1 N H2SO4 as diluent. Mix thoroughly by inversion.

> Sulfuric Acid: 0.1 N. Add 9.0 mL dilute H2SO4 (6 N) to 500 mL volumetric flask and dilute to the mark with distilled water. Mix.

> Extraction Solvent: Mix 3 volumes of dichloromethane with 1 volume of isopropanol. Mix thoroughly. Do not make any more than you need. (4 mL solvent/test.)

> pH 8.5, 0.1 M Ammonia Buffer: Dissolve 5.4 g NH4Cl in 90 mL (use a 100 mL vol. flask) distilled water, then add 2.5 mL 6 N NH4OH. Adjust to pH 8.5 using 6 N NH4OH or 6 N HCl. Dilute to mark with distilled water and mix.

Procedure: Label a series of 16 × 125 mm screw-cap test tubes as: Blank, STD-5, STD-10, STD-15, sample/s, Control. Add 2.0 mL of appropriate sample to each tube, using distilled water for the BLANK. Then add 1.0 mL ammonia buffer to each of the six tubes and mix. Add 4.0 mL of extraction solvent to all tubes, caps, and mix by inversion for 5-10 minutes. Centrifuge the tubes for 2 minutes at 500 RPM using a clinical centrifuge. Use a setting of 8 on "Speed Control." Aspirate the (UPPER) aqueous layers and then decant each (LOWER) organic layer into a clean, correspondingly labeled 16 x 125 mm screw-cap tubes. Add 4.0 mL of 0.1 N H2SO4 to each tube. Cap tubes, and mix by inversion for 5-10 minutes. Centrifuge tubes for 2 minutes and 1500 RPM. Transfer each (UPPER) aqueous layer to a labeled test tube by means of a Pasteur pipet. Start the spectrofluorometer, following the instructions as listed in the instruction manual. Consult the instruction manual for guidelines on setting slit widths and scan speeds.

Emission spectra of quinine in 0.1 N H2SO4: Use 10 mg/L standard as originally prepared. Set the excitation wavelength to 350 nm and scan from 200 nm to 750 nm using the emission monochromator.

Excitation Spectra of quinine in 0.1 N H2SO4: Using the same solution, and an emission wavelength near the maximum found from the above step, scan the excitation wavelength from 200-340 nm.

Set the excitation wavelength to 350 nm and perform an emission scan of blank solution from 300-700 nm. Set the excitation wavelength to 350 nm. Set emission wavelength to the optimum wavelength found from the emission scan. Place 15 mg/L extracted standard in the cuvette and adjust controls to just keep recorder pen on scale. Now run an emission scan from 300-600 nm. With controls adjusted, run emission scans on all remaining solutions including control and unknown.

Reserpine	**Utilization:**Reserpine is an indole alkaloid obtained from the roots of *Rauvolfia serpentina*, *R. micmntha*, *R. vomi-form* and *R. tetraphylla* of

Contd...

family *Apocynaceae*. Reserpine is used as potent hypotensive and sedative agent. Reserpine irreversibly blocks the vesicular monoamine transporter (VMAT) and causes depletion of intracellular norepinephrine, serotonin, and dopamine. It may take days to weeks to replenish the depleted VMAT, and thus, reserpine's effects are long-lasting. Depletion of dopamine can lead to drug-induced parkinsonism too.

IUPAC name: methyl (3β,16β,17α,18β,20α)-11,17-dimethoxy-18-[(3,4,5-trimethoxybenzoyl)oxy]yohimban-16-carboxylate

Isolation: Take the accurately weighed quantity of dried rauvolfia powder to obtain coarse sized particles. Extract the powder with 90% alcohol in soxhlet apparatus. Filter the alcoholic extract and evaporate to dryness. Extract the residue with the mixture of ether: chloroform: 90% alcohol (20:8:2.5). Filter and add dilute ammonia to filtrate. Add water to precipitate the crude alkaloid mixture. Collect the precipitate and dry. Dissolve the residue in 0.5N H_2SO_4. Add ammonia to the above solution to liberate alkaloid salts. Extract the solution with 3 portions of chloroform. Evaporate the chloroform to obtain total rauvolfia alkaloids. Separate the reserpine from the crude extract by column chromatography separation technique.

Identification: Take sample to this add ethanol and vanillin in acetic acid. Violet-red color indicates presence of reserpine.

Thin Layer Chromatography (TLC): Stationary phase: Silica gel G, **Mobile phase: Chloroform:Acetone:** Diethyl ether (50:40:10), Detecting agent :Dragendorff's reagent, Rf Value : 0.72-0.35 Orange spot

Estimation Method: Bromocresol green Reagent method

Curcumin	**Utilization:** Curcumin is diarylheptanoid and principal curcuminoid present in the rhizomes of *Curcuma longa* offamily Zingiberaceae commonly called as turmeric. This is used as an anti inflammatory, antiseptic, anticancer and hepatoprotective agent. It is also a common food colourant. Curcumin is a bright-yellow color and may be used as a food coloring. As a food additive, its E number is E100.

Contd...

Curcumin

demethoxycurcumin

bisdemethoxycurcumin

Curcumin IUPAC name: (1E,6E)-1,7-Bis(4-hydroxy-3-methoxyphenyl)-1,6-heptadiene-3,5-dione

Isolation: Take accurately weighed quantity of turmeric powder. Extract with n-hexane for 2 hrs. Discard the n-hexane extract and extract marc with methanol for 2 hrs. Distill off the methanol to obtain solid residue. Dissolve this residue in any suitable alkali, filter and acidify with HCl to precipitate crystals of curcumin.

Identification: Treat sample with acetic anhydride and conc. H2SO4, it gives violet colour. Then observe under UV light, red fluorescence confirms presence of curcuminoids.

Estimation by HPTLC: Stationary phase-silica gel 60 F254, Mobile phase- chloroform: methanol (9.5:0.5), Detection: 1 % alcoholic KOH solution, TLC scanner wavelength 421 nm.

4.1 Forskolin

Utilization: Forskolin is a labdane diterpenoid present in Indian coleus plant (*Coleus forskohli*, Family Lamiaceae) have antihypertensive, positive inotropic, platelet aggregation inhibitory and adenylate cyclase activating properties.

IUPAC name: (3R,4aR,5S,6S,6aS,10S,10aR,10bS)-6,10,10b-trihydroxy-3,4a,7,7,10a-pentamethyl-1-oxo-3-vinyldodecahydro-1H-benzo[f]chromen-5-yl acetate

Isolation: Take weighed powder of *C. forskolii* tubers. Extract it with methanol. Filter and concentrate methanol extract. Dissolve methanol extract in chloroform. Take this chloroform extract in separating funnel

Contd...

	andadd water. Shake slowly and separate water layer. Repeat three times. Now evaporate purified chloroform layer to $1/4^{th}$ volume. Add ice cold n-hexane to precipitate Forskolin. **Estimation: HPLC Method** **Stationary Phase/Column**: C18, **Mobile phase:** Water, Acetonitrile (Gradient), flow rate of 1 mL/min, Detection: 210 nm
4.2 Sennoside	**Utilization:**Sennosides are obtained from dried leaflets of *Cassia auriculata* (Alexandrian senna) and *Cassia angustifolia* (Tinnevelly senna); it belongs to the family *Leguminaceae*. Sennoside A, B, C, and D are isomers of each other. These are useful as potent laxatives. Sennosid A: R = COOH Sennosid C: R = CH₂OH Sennosid B: R = COOH Sennosid D: R = CH₂OH IUPAC Name: Sennoside-A (9R)-9-[(9R)-2-carboxy-4-hydroxy-10-oxo-5-[(2S,3R,4S,5S,6R)-3,4,5-trihydroxy-6-(hydroxymethyl)oxan-2-yl]oxy-9H-anthracen-9-yl]-4-hydroxy-10-oxo-5-[(2S,3R,4S,5S,6R)-3,4,5-trihydroxy-6-(hydroxymethyl)oxan-2-yl]oxy-9H-anthracene-2-carboxylic acid **Isolation:**Take the accurately weighed quantity of coarse powder of *Senna* leaves. Extract the powder with benzene on an electric shaker for 2 hours. Filter and discard the benzene extract. Extract the dried marc with 70% methanol for 5 hours. Filter and again extract the marc with fresh methanol for 2 hours. Filter and combine both extracts and concentrate to $1/4^{th}$ of its original volume. Add sufficient quantity of HCl to obtain a pH of 3.2 and keep aside for 2 hours at 5°C. Filter and add sufficient quantity of alcoholic anhydrous calcium chloride with continuous stirring. Add ammonia to bring the pH to 8 and keep aside for 2 hours. Filter and collect the precipitates of calcium sennoside. Dry the precipitate in a desiccator and calculate the percentage yield. **Estimation methods: (Magnesium acetate Method)**

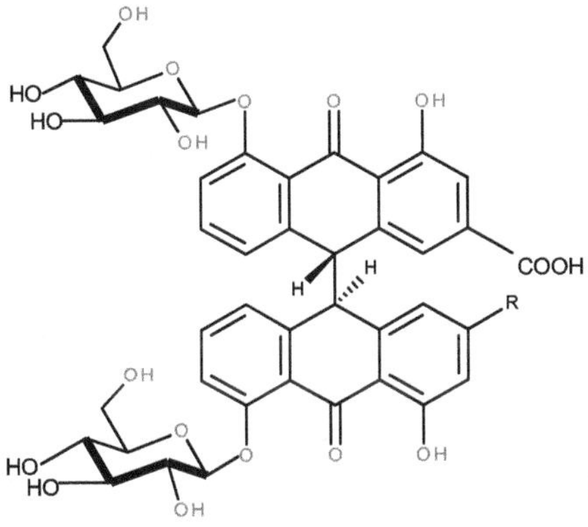

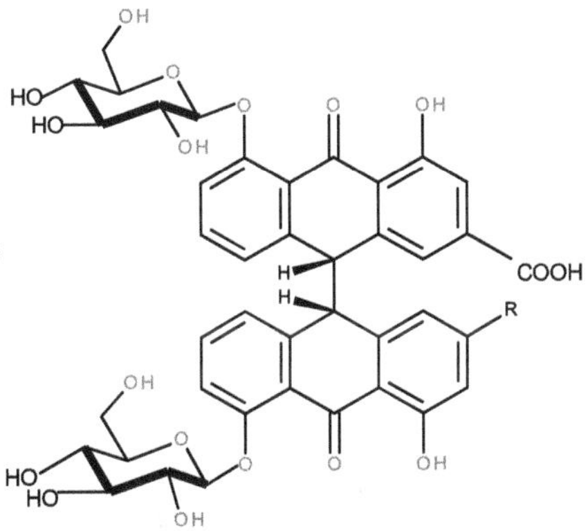

Contd...

	Take dried extract of senna or isolated sennoside, dissolve it in 10 ml of 0.5% w/v magnesium acetate in methanol yielding a red solution. Measure the absorbance at 515 nm.Prepare five different concentration solutions of standard anthraquinone (Sennoside A, B, C or D) in Beer-Lambert's law obeying range of concentration. Prepare calibration curve and calculate the percentage of total anthraquinone present in the sample/s.
4.3 Artemisinin	**Utilization:**Artemisinin is sesquiterpenoid lactone present in the flower heads of *Artemisia annua* of family *Asteraceae*. It is an antimalarial drug found to be effective against chloroquine resistant strains of *P. falciparum*. IUPAC name: (3R,5aS,6R,8aS,9R,12S,12aR)-Octahydro-3,6,9-trimethyl-3,12-epoxy-12H-pyrano[4,3-j]-1,2-benzodioxepin-10(3H)-one **Isolation**: Take an accurately weighed quantity of *Artemisia* leaves powder. Extract the powder with petroleum ether. Filter and evaporate the filtrate to dryness. Dissolve the residue in chloroform. Add acetonitrile to precipitate the impurities. Separate the precipitate and discard. Concentrate the filtrate to dryness. Load this residue on a column chromatography by using chloroform-ethyl acetate mixture. Monitor the elution by TLC. Combine the fraction containing artemisinin. Purify the crude artemisinin by using aqueous alcohol. **Identification:**Boil Artemisinin in 10ml of alcohol and filter. Add Sodium hydroxide to filtrate and heat. Red colour will indicate presence of Artemisinin. **Estimation: TLC Densitometry Method** **Stationary Phase:** Silica Gel G, **Mobile Phase:** toluene: ethyl acetate (10:1), **Detection:** Vanillin Sulfuric acid Reagent, **Detection wavelength:** 520 nm
4.4 Diosgenin	**Utilization:**Diosgenin is obtained from the dried tubers of *Dioscoreadeltoidea*and other species of *Dioscorea* (Dioscoreaceae). It is useful in the synthesis of vitamins and sex hormones. The sugar-free (aglycone) diosgenin is used for the commercial synthesis of cortisone, pregnenolone, progesterone, and other steroid products.

Contd...

DioscinDiosgenin

Diosgenin IUPAC name: (3β,25*R*)-spirost-5-en-3-ol

Isolation: Take the accurately weighed quantity of coarse powder of *Dioscorea* tubers. Extract the powder with methanol for 6 hours. Filter and concentrate the filtrate to obtain a semisolid mass. Add dilute HCl and stir for 6 hours. Precipitates of diosgenin will be observed. Filter and purify the precipitate with alcohol.

Identification Salkowski Reaction: Dissolve 1–2 mg of the sample in 1 ml of CHCl3 and add 1 ml concentrated H2SO4. Chloroform layer shows red color and acid layer shows green fluorescence

Estimation Method: Liberman-Burchard method

Make different concentrations from diosgenin working standard (10-100 µg/ml in chloroform). To it add 2 ml Liberman-Burchard reagent. Keep it for 10-15 min in cold water.Read the absorbance at 620 nm. Follow the same procedure for extract samples. Calculate the amount of total steroids from calibration curve as a diosgenin equivalent.

4.5 Digoxin	**Utilization:**Digoxin is a potent cardioactive glycoside obtained from the leaves *of Digitalis purpurea* and *D. lanata*of the family *Scrophulariaceae*. Digoxin is widely used in the treatment of various heart conditions, namely atrial fibrillation, atrial flutter, and sometimes heart failure that cannot be controlled by other medication. D-Glucose : digoxin = Digoxigenin + 3 Digitoxose sugar IUPAC name: 4-[(3S,5R,8R,9S,10S,12R,13S,14S)-3-[(2S,4S,5R,6R)-5-[(2S,4S,5R,6R)-5-[(2S,4S,5R,6R)-4,5-dihydroxy-6-methyl-oxan-2-yl]oxy-4-hydroxy-6-methyl-oxan-2-yl]oxy-4-hydroxy-6-methyl-oxan-2-yl]oxy-12,14-dihydroxy-10,13-dimethyl-1,2,3,4,5,6,7,8,9,11,12,15,16,17-tetradecahydrocyclopenta[a]phenanthren-17-yl]-5H-furan-2-one **Isolation**: Take accurately weighed quantity of fresh lea ith

Contd...

	petroleum ether. Grind this defatted leaves with neutral salt to inactivate the enzymes. Extract the pulp with ethyl acetate for 2 hours. Filter and concentrate the filtrate to a dry residue. Column chromatographic separation of the above residue yields lanatoside A (46%), lanatoside B (17%) and lanatoside C (37%). Take lanatoside C and treat with dilute HCl to get a hydrolysis product i.e. digoxin. **Identification:** Legal Test (Cardenolides): Mix 1 ml of test solution with 2 ml pyridine and sodium nitroprusside. Pink or red color confirms presence of digoxin. **Estimation method: DNBA Method** Prepare different dilutions of standard digitoxin in water. Add 5ml lead acetate (15%) and 7.5 ml disodium hydrogen phosphate (4%). Mix and filter. To the filtrate, add 5 ml of 15% HCl and reflux for 1 hour. Then, extract this solution with 3 portions of 25 ml chloroform. Separate and combine chloroform layer. Evaporate the chloroform extract up to 50 ml and add 7 ml 50% ethanol.Then add 2 ml of 3,5-Dinitrobenzoic acid (DNBA) and 1 ml of 1M NaOH. Mix well and measure the absorbance at 540 nm. Repeat the same procedure as mentioned above for extracted sample/s.Then, measure the absorbance and calculate percentage quantity from the calibration curve of standard digitoxin.
4.6 Atropine	**Utilization:**Atropine is a tropane alkaloid obtained from *Atropa belladonna* (deadly night shade), *Datura stramonium* (thorn apple), and *Hyoscyamus niger*(henbane) of family *Solanaceae*. It is found in many members of the *Solanaceae* family. It is a diastereomeric mixture of d-hyoscyamine and l-hyoscyamine with most of its physiological effects due to l-hyoscyamine. Its pharmacological effects are due to binding to muscarinic acetylcholine receptors. It is an antimuscarinic and anticholinergic agent. It is mydriatic and antidote in opium poisoning. Atropine is contraindicated in patients pre-disposed to narrow angle glaucoma. IUPAC name: [(1R,5S)-8-methyl-8-azabicyclo[3.2.1]octan-3-yl] 3-hydroxy-2-phenylpropanoate **Isolation**: Take the accurate weighed quantity of powder of belladonna roots or datura leaves. Extract the powder with 95% ethanol. Filter it and add potassium carbonate to filtrate to convert hyoscyamine to atropine. Now, add chloroform to this solution. Shake well and separate chloroform layer. Evaporate chloroform to dryness. Diss n

Contd...

	dilute H_2SO_4.Add potassium carbonate to precipitate atropine. Separate precipitate and extract with ether. Add oxalic acid to ether solution to give atropine oxalate crystals. **Identification: Vitali Morin Test:** To the extract solution, add few fragments of magnesium ribbon and concentrated HCl drop wise, yellowish; yellow- orange occasionally orange colour appears after few minutes **Estimation Method: (Vitali-Morin Method)** Prepare 5 test tubes containing increasing quantity of atropine working standard from the range of 2000-1000µg/ml.To it, add 0.2 ml fuming nitric acid. Mix and evaporate to dryness. To this residue, add 2 ml acetone and 0.1 ml of 3% methanolic KOH. Make up the volume to 10 ml with acetone. Read the absorbance at 430 nm or by using a green filter. Follow the same procedure for the crude alkaloid containing chloroform extract from plant samples. Calculate the amount of total alkaloids from calibration curve of atropine or L- Hyoscyamine sulphate as standard.
4.7 Podophyllotoxin	**Utilization:**Podophyllotoxin is the lactone resin present in the root and rhizome of *Podophyllum hexandrum*belonging tothe family*Berberidaceae*. It is used as an antiproliferative/ anticancer agent. Synthetic derivatives include etoposide, teniposide, and etopophos. 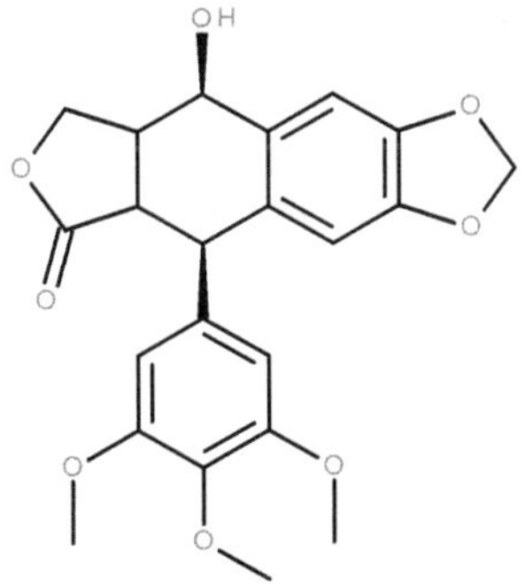 Podophyllotoxin 2 IUPAC name: (10R,11R,15R,16R)-16-hydroxy-10-(3,4,5-trimethoxyphenyl)-4,6,13-trioxatetracyclo[7.7.0.03,7.011,15]hexadeca-1,3(7),8-trien-12-one **Isolation:** Take accurately weighed quantity of rhizome/roots of *Paeonia emodi*with methanol. Filter and evaporate to semisolid mass. Dissolve semisolid mass into acidic water. Precipitate is formed which should be allowed to settle for at least 2 hours.Filter and wash the filtrate with cold water. Collect the residue, wash with acidified water, and dry to obtain dark brown amorphous powder. Extract the residue with hot alcohol. Filter and evaporate to dryness. Recr *Contd...*

residue in benzene to yield podophyllotoxin.

Identification: Treat podophyllotoxin with 50% Sulphuric acid. It will show violet-blue colour

Thin Layer Chromatography (TLC): Stationary phase: Silica gel G, Mobile phase: Chloroform: Methanol (90:10) for about 6cm (Only glycosides are separated but aglycone like podophyllotoxin remains in the region of the front. The same plate is again eluted with more weakly polar Solvent Chloroform: Acetone (65:35) upto 12cm. Detecting agent: Spray with methanol Sulphuric acid and heat 10 minutes at 110^0 C. RF Value : 0.65 Yellow spot

Estimation:TLC Densitometry Method

Stationary Phase: RP-18 F254 TLC plates, **Mobile Phase:** acetonitrile–water (4:6, v/v) ,**Detection:** UV Detector at 217 nm, Rf Value: 0.42

4.8 Caffeine

Utilization: Most of the people start their morning with the cup of tea which contains the alkaloid caffeine. Caffeine is a naturally occurring substance found in the leaves, seeds or fruits of over 63 plant species worldwide, and is part of a group of compounds known as methylxanthines. The most commonly known sources of caffeine are coffee, cocoa beans, cola nuts, and tea leaves. Caffeine is a bitter, white crystalline xanthine alkaloid that is a psychoactive stimulant drug. Caffeine is pseudo alkaloid as it is not biosynthesized from amino acids, but gives all alkaloid identification tests positive. Extraction of caffeine is a simple decoction process where caffeine is first dissolved in water and separated by chloroform. Tea leaves contain 1%-4% of caffeine while coffee seeds contain 1%-2% of caffeine.

IUPAC name: 1,3,7-trimethylpurine-2,6-dione

Isolation: Weigh 10 g of tea leaves and transfer to 250 ml distilled water. Boil the water for 30 minutes with occasional stirring. Then allow to cool andfilter the solution. Take filtrate in a separating funnel and to it, add 100 ml chloroform. Shake vigorously so that total caffeine will be transferred to chloroform. Separate chloroform layer. Evaporate chloroform over water bath. White caffeine crystals will collect at the bottom.

Identification:

Contd...

	*Murexide test:*Treat caffeine powder with HCl and KCl, heat to dryness, and expose residue to dilute ammonia to get purple color. *Tanniac acid test:* Treat caffeine powder with tannic acid solution to produce a white precipitate. **Thin Layer Chromatography (TLC):** Stationary phase : Silica gel-G, Mobile phase : Ethyl acetate: methanol: acetic acid (80:10:10), Detecting agent : Expose to vapours of iodine, RF Value : 0.41 Brown spot **Estimation Method:** HPLC Caffeine solution:50-1000 ppm, Column:C18, Mobile phase: Methanol :water (50:50), Flow rate:1 ml/min, Detection:254 nm
4.9 Taxol	**Utilization**: Taxol is a diterpenoid alkaloid obtained from bark of plant *Taxus brevifolia* of family *Taxaceae*. It is a strong antineoplastic agent. Paclitaxel is a medication used to treat a number of types of cancer including ovarian cancer, breast cancer, lung cancer, and pancreatic cancer as well as Kaposi's sarcoma. Paclitaxel's mechanism of action involves interference with the normal breakdown of microtubules during cell division. Paclitaxel is mentioned in the World Health Organization's List of Essential Medicines. Paclitaxel (IUPAC) name: (2α,4α,5β,7β,10β,13α)-4,10-Bis(acetyloxy)-13-{[(2R,3S)-3-(benzoylamino)-2-hydroxy-3-phenylpropanoyl]oxy}-1,7-dihydroxy-9-oxo-5,20-epoxytax-11-en-2-yl benzoate **Isolation:**Take the accurately weighed quantity of powder of bark of taxus plant. Extract the powder exhaustively with ethanol or methanol. Filter and evaporate filtrate at 40^0C to dryness. Dissolve this residue again in methanol. Filter and evaporate filtrate to dryness. Now dissolve this residue in a mixture of carbon tetrachloride and water. Filter and centrifuge. Separate and combine carbon tetrachloride layer. Evaporate this layer to dryness. Dissolve this residue in mixture of methanol and carbon tetrachloride (1:1). Filter and evaporate to obtain dry residue of taxol alkaloids. This crude residue can be further purified by preparative TLC using silica gel as stationary phase and mobile phase of carbon tetrachloride and methanol (95:5).(Rf value of taxol is $0.35 \approx 0.37$).

Contd...

	Identification: **Estimation:** UV Spectrophotometric method: Standard Paclitaxel solution: 100µg/ml, Solvent: methanol:Phosphate buffer (30:70) solution, Absorabance wavelength: 230 nm. Prepare calibration curve of standard Taxol and estimate concentration of unknown sample from it.
4.10 Vincristine and Vinblastine	**Utilization:**Vinblastine and vincristine are indole alkaloids obtained from the whole plant of *Catharanthus roseus Linn* of family *Apocynaceae*. Both are used as antineoplastic agent in Hodgkin's lymphoma, non-small cell lung cancer, bladder cancer, brain cancer, and testicular cancer. It is mentioned in the World Health Organization's List of Essential Medicines. Vincristine Vinblastine **Vincristine (IUPAC) name:** (3aR,3a1R, 4R, 5S,5aR,10bR)-methyl 4-acetoxy-3a-ethyl-9-((5S,7S,9S)-5-ethyl-5-hydroxy-9-(methoxy carbonyl)-2,4,5,6,7,8,9,10-octahydro-1H-,7methano[1]azacycloundecino[5,4b]indol-9-yl)-6-formyl-5-hydroxy-8-methoxy-3a,3a1,4,5,5a,6,11,12-octahydro-H-indolizino[8,1-cd]carbazole-5-carboxylate **Vinblastine (IUPAC) name:**dimethyl(2β,3β,4β,5α,12β,19α)-15-[(5S,9S)-5-ethyl-5-hydroxy-9-(methoxycarbonyl)-1,4,5,6,7,8,9,10 octahydro-2H-3,7-methanoazacycloundecino [5,4-b] indol-9-yl]-3-hydroxy-16-methoxy-1-methyl-6,7-didehydroaspido spermidine-3,4 dicarboxylate **Isolation:** Take the accurately weighed quantity of dried entire *Vinca* plant and powder to obtain coarse sized particles. Extract the powder with the mixture solution of alcohol-water-acetic acid in a ratio of 9:1.Filter and evaporate the filtrate to dryness. Dissolve the residue in 2% hot HCl. Filter and adjust the pHof filtrate to 7.0. Extract the neutralized filtrate with benzene and separate benzene layer. Evaporate the benzene layer to give crude alkaloidal residue. Subject the dried residue to column chromatography containing alumina adsorbent. Use the gradient elution technique starting from pure benzene, then benzene: chloroform (1:1), and lastly chloroform: methanol (1:1) mixture.

Contd...

<table>
<tr><td></td><td>

Monitor the column fractions with TLC [Stationary Phase: silica gel, Mobile Phase: ethyl acetate: methanol (9:1)] Collect and combine the fractions giving single spot of vinblastine. Further, elution of the column results in separation of vincristine.

Estimation Method:

Bromocresol green (BCG) Reagent method: Take 0.4. 0.6, 0.8, 1.0, and 1.2 ml atropine solution in a separate test tube.Add 5 ml of Phosphate Buffer Solution (pH 4.7) and 5 ml of BCG (Dissolve 69. 8 mg BCG in 3 ml 2N NaOH and 5 ml distilled water. Make up the volume up to 100 ml) solution. Shake well and extract the yellow colored complex with chloroform. Separate chloroform and make up the volume to 10 ml. Measure the absorbance at 470 nm against blank. Now prepare the methanolic extract of plant material. Dry and dissolve in 2N HCl. Filter and wash with chloroform. Adjust the pH neutral with 0.1 N NaOH.Now add 5ml of Phosphate Buffer Solution (pH 4.7) and 5 ml of BCG solution. Shake well and extract the yellow colored complex with chloroform. Separate chloroform and make up the volume to 10 ml and measure the absorbance at 470 nm.Calculate concentration of total alkaloids from calibration curve of atropine standard.

HPLC Method:Column: C18, **Mobile phase:** 5%–95% acetonitrile in water with 0.01% trifluoroacetic acid, Flow Rate:0.5 ml/min, Detection: 220 and 254 nm

</td></tr>
</table>

Subjective Questions

1. How to isolate, estimate and utilize sennosides or caffeine?

2. How to isolate, estimate and utilize Forskolin or digoxin?

3. How to isolate, estimate and utilize taxol or podophyllotoxin?

4. How to isolate, estimate and utilize Vincristine and vinblastine?

5. How to isolate, estimate and utilize Artemisinin or diosgenin?

6. What are commercial uses of Forskolin, diosgenin and Artemisinin?

Multiple Choice Questions (MCQs)

1. Forskolin was first discovered by
   a. Hoeschst Research Centre, Mumbai
   b. Central Drug Research Institute (CDRI), Lucknow, India
   c. Both a and b
   d. None of the above

2. Forskolin inhibits human platelet aggregation induced by

    a. Epinephrene

    b. Collagen

    c. Both a and b

    d. None of the above

3. Forskolin production can be enhanced at industrial level by implementing

    a. Transformed cell suspension culture

    b. Molecular cloning

    c. Transformed hairy root culture

    d. All of the above

4. The export of calcium sennosides during 1995-96 and 96-97 was approximately

    a. Rs. 194 lac and Rs. 330 lac respectively

    b. Rs. 195 lac and Rs. 340 lac respectively

    c. Rs. 196 lac and Rs. 350 lac respectively

    d. Rs. 197 lac and Rs. 360 lac respectively

5. Which strain of yeast is used for fermentative production of artemisinic acid and its chemical conversion to artemisinin at industrial level?

    a. Saccharomyces cerevisiae

    b. Saccharomyces pastorianus

    c. Saccharomyces paradoxus

    d. None of the above

6. For estimation of isolated artemisinin by TLC, which mobile phase composition is suitable?

    a. Petroleum ether : ethyl acetate (1:2)

    b. Chloroform: ethyl acetate (1:2)

    c. Toluene : ethyl acetate (1: 2)

    d. None of the above

7. Artimisinin is estimated by using HPLC at UV wavelength

    a. 360 nm

    b. 260 nm

    c. 160 nm

    d. 200 nm

8. Artemisinin is used commercially against chloroquine resistant strains of

    a. Plasmodium vivax

    b. Plasmodium falciparum

    c. Both a and b

    d. None of the above

9. Which variety of Dioscorea is used as raw material in industry for production of diosgenin?

    a. Dioscoreadeltoididea

    b. Discoreacomposite

    c. Dioscoreazingiberansis

    d. Dioscoreaflouribunda

10. Which method can provide fast diosgenin level enhancement at industrial level?

    a. Hairy root culture

    b. Bioreactors

    c. Large scale cultivation

    d. callus culture

11. Digoxin tablet USP are indicated for treatment of
    a. Severe heart failure in adults
    b. Mild to moderate heart failure in adults
    c. Mild heart failure in adults
    d. None

12. Digoxin is estimated by UV spectrophotometry at wavelength
    a. 320 nm
    b. 220 nm
    c. 520 nm
    d. 420 nm

13. Atropine is estimated by UV spectrophotometry at UV wavelength
    a. 252 – 262 nm
    b. 352 – 362 nm
    c. 152 – 162 nm
    d. 452 – 462 nm

14. Atropine is administered as
    a. 1 ml – 1.5 ml in the form of belladonna tincture – 2 times a day
    b. 0.6 – 1ml in the form of belladonna tincture – 4 times a day
    c. 2 ml in the form of belladonna tincture – 3 times a day
    d. 5 ml in the form of belladonna tincture – 2 times a day

15. HPLC estimation parameters for podophyllotoxin are
    a. Mobile phase – methanol: water (62 : 38 v/v); detector wavelength 280 nm
    b. Mobile phase – methanol: water (62 : 48 v/v); detector wavelength 380 nm
    c. Mobile phase – methanol: water (62 : 58 v/v); detector wavelength 480 nm
    d. Mobile phase – methanol: water (62 : 38 v/v); detector wavelength 180 nm

16. Current worldwide production of caffeine is
    a. Larger than 5.5 million tons
    b. Less than 5.5 million tons
    c. Equal to 5.5 million tons
    d. None of these

17. Decaffeinated coffee is prepared by removing
    a. 100% caffeine
    b. More than 97% caffeine
    c. More than 90% caffeine
    d. More than 95% caffeine

18. Vincristine and vinblastine are costlier commercially because
    a. Extreme low content of alkaloids
    b. High amount of alkaloids
    c. Special nutrients are required
    d. Endangered plant

19. Production of taxol can be enhanced by

    a. Chemical synthesis

    b. Endophytic fungi

    c. Cultivation on large area

    d. Both a and b

Answer Key

1.c	2. c	3. d	4. a	5. a	6.a	7.a	8. b	9. c	10. a
11. b	12. b	13. a	14. b	15. a	16. a	17. b	18. a	19.d	

Unit 5

Basics of Phytochemistry

PCI Syllabus

Metabolic pathways in higher plants and their determination

- Brief study of basic metabolic pathways and formation of different secondary metabolites through these pathways- Shikimic acid pathway, Acetate pathways and Amino acid pathway.

- Study of utilization of radioactive isotopes in the investigation of Biogenetic studies

Chapter Content

5.1   Basics of Phytochemistry

5.2   Methods of Extraction: Traditional and Modern

5.3   Application of latest techniques in the isolation, Purification and identification of Crude drugs

    5.3.1 Chromatography

        5.3.1.1   Planer Chromatography

        5.3.1.2   Column Chromatography

    5.3.2 Electrophoresis

    5.3.3 Spectroscopy

5.1 Basics of Phytochemistry

Phytochemistry is the study of phytochemicals or chemicals derived from plants through primary or secondary metabolism. Common phytochemical techniques used in the field of phytochemistry are extraction, isolation, separation by chromatography or electrophoresis techniques and structural elucidation by spectroscopic techniques.

5.2 Methods of Extraction: Traditional and Modern

Traditional methods of Extraction

Maceration	As per IP 1966 maceration is: "Place the solid material with the whole of the menstruum in a closed vessel and allow it to stand for 7 days, shaking occasionally, and strain, pressing the marc and mixing the liquids obtained. Clarify by subsidence or filtration". Seven days are considered to be an adequate period of time to bring about equilibrium between solute and solvent. Closed vessel is preferred to avoid undue loss of solvent due to evaporation and contamination.For water maceration, add 5% chloroform to avoid fungal growth.
Percolation	As per IP 1966 percolation is carried out as follows: "Moisten the solid materials with sufficient quantity of the menstruum, allow it to stand for 4 hours in a well closed vessel, packed in a percolator, and add sufficient of menstruum to saturate the material. When the liquid commences to drop from the percolator, close the outlet, add sufficient amount of the menstrum to leave a layer above the drug and allow it to stand for 24 hours. Allow percolation to proceed slowly until the percolate measures about three quarters of the volume required for the finished tincture. Press the marc, mix the expressed liquid with the percolate, and sufficient of menstrum to produce the required volume. Clarify by subsidence or filtration.
Hot continuous counter-current extraction [Soxhlet extraction]	This is also called as Soxhlet extraction due to specially designed apparatus. It is the process where the same quantity of solvent is made to circulate through the extractor of drug by evaporation and subsequent condensation. With less solvent complete extraction is possible. But this method is not suitable for thermolabile components.

Contd...

Infusion	This is a very simple method of extraction used for vitamins, volatile ingredients and soft ingredients in which the powdered drug is extracted with hot or cold water.
Decoction	The word decoction means to concentrate by boiling. In this method, the drug powder can be boiled with water for a few minutes to hours and then filtered.
Distillation (For volatile oil)	This method is useful for extraction of essential oils. It involves the heating of plant material with water in a distillation unit. The vapours of the volatile oil component are passed towards the condenser along with the steam vapours. The oil and water layer separate on cooling.
Expression (For volatile oil)	It is a purely mechanical technique which involves sponge, ecuelle and mechanical methods to extract essential oils. Volatile oil absorption on sponge by rupturing the oil glands from citrus peels by squeezing is the best example of sponge method. In Ecuelle method, the sharp projection containing vessel is used to rupture the glands from the citrus peels.
Enfleurage (For volatile oil)	This method is useful for extraction of essential oils from delicate plant parts like rose oil from rose petals. Fatty material is spread evenly into the glass plate as a thin layer. The petals or inflorescences are spread on this fatty layer. The material is allowed to extract completely in fat by keeping for 24 hours. Then, again fresh petals are loaded. This process continues till the fatty layer is saturated with volatile oil.
Couther current distribution (CCD)	This is a liquid-liquid extraction process and is based on the principle of partition coefficient i.e. the liquid to be isolated is much more soluble in either one of the two immiscible liquids. The apparatus **(Craig Apparatus)** consists of a series of

Contd...

	interconnected tubes in order to repeat the same cycle of process for efficient separation. The first tube in the series contains three liquids • Heavy stationary phase liquid • Mixture to be separated • Light mobile phase liquid.
Droplet Counter Current Chromatography (DCCC)	DCCC is based on the combined principle of liquid-liquid extraction and column chromatography. It is a simpler modification of counter current distribution (CCD) principle of Craig apparatus. Here, the mobile phase moves in the form of small droplets or plates. Hence the solute separates by distributing between stationary and droplets according to the partition coefficient differences. Stationary and mobile phase moves in the counter current direction and hence it is called as counter current chromatography.
Solid-Phase Extraction (SPE)	Solid-phase extraction (SPE) is a modern chromatographic separation technique. It separates the individual components from the liquid mixture in accordance with their physical and chemical properties. It is of three types i.e. normal, reversed or ion exchange SPE. Unlike liquid-liquid extraction, it doesn't form emulsion. SPE uses the principal of chromatographic separation.
Liquid-liquid Extraction (LLE)	Liquid-liquid extraction or solvent extraction is the most preferred extraction/ separation method of choice in organic chemistry. This process is based on partition coefficient that allows the separation of components of mixture due to their unequal solubility in two immiscible liquid phases.
Modern methods of extraction	
Ultrasound Extraction	It is also called as sonication extraction. In this method, sound wavesof high-frequency pulses of 20 kHz are generated in an ultrasonic bath. The samples are kept in a suitable vessel containing appropriate solvent. The waves generated from the transmitter easily penetrate the cell membrane byinducing a mechanical stress on the cells. This increases cell

Contd...

Microwave Assisted Extraction	Microwaves (frequency 300 MHz to 300 GHz) are non-ionising electromagnetic waves which are useful in the extraction of phytoconstituents. The principles on which microwave works is that the high temperature produced by microwaves evaporates the moisture present in the plant powder which dehydrates cellulose, ruptures the cell walls and thus solubilises phytoconstituents. Rapid exhaustive extraction within 5-10 min.
Accelerated Solvent Extraction (ASE)	It is also known as pressurised solvent extraction (PSE) and is very similar to a Soxhlet extraction. This is a relatively new fully automated technique which operates at high temperatures and pressures (100–200 bars) ,but at the same time, keeps the solvents in liquid form during the extraction process. The solvents used are near their supercritical state where they have high extraction properties. It is very similar to supercritical fluid extraction, the only difference is the pressure is high but below the supercritical point and ACE does not involve the use of carbon dioxide.
Supercritical Fluid Extraction (SCF)	This is organic solvent free extraction method. At super critical point (temperature and pressure above critical point) gas converts to fluid. Such super critical fluid solvent mediated extraction is a process, very similar to simple extraction where solvent passes through the coarsely powdered sample and extracts its constituents. The only difference here is that there is a special assembly to control temperature and pressure. It is easy to remove the solvent extract because carbon dioxide escapes as gas as the pressure is released. Gas can be re-cycled and reused.

5.3 Application of latest Techniques in the Isolation,Purification and Identification of Crude Drugs

5.3.1 Chromatography

Definitions: It is a technique of separating a mixture of components by distributing between two phases (stationary phase and mobile phase) depending on the relative affinities of the components for the two phases. Stationary phase may be solid, gel, or liquid supported by solid. Mobile phase may be liquid or gas.

Types/Classification

Classification based on technique of separation

- *Planar chromatography*: Paper, Thin layer chromatography (TLC), High performance thin layer chromatography (HPTLC)

- *Column chromatography*: Open column chromatography, gas chromatography (GC), High pressure liquid chromatography (HPLC), Vacuum liquid chromatography (VLC), Low pressure liquid chromatography (LPLC), Ultra pressure liquid chromatography (UPLC), Ion exchange chromatography (IEC), Gel chromatography, Flash chromatography.

Classification based on principle of separation

➤ **Adsorbtion** is purely surface phenophenon where separation of components in a mixture introduced into chromatography system is based on the relative adsorption affinities of components to the stationary phase.

➤ **Partition** chromatography is process of separation whereby the components of the mixture get distributed into two liquid phases due to differences in partition coefficients.

➤ **Ion-exchange chromatography** (or ion chromatography) is a process that allows the separation of ions and polar molecules based on their affinity to the ion exchanger.

➤ **Size-exclusion** chromatography (SEC) is a chromatographic method in which molecules in solution are separated by their size, and in some cases molecular weight. Small molecule will elute late and large molecule that cannot penetrate any region of the pore system will elute earlier.

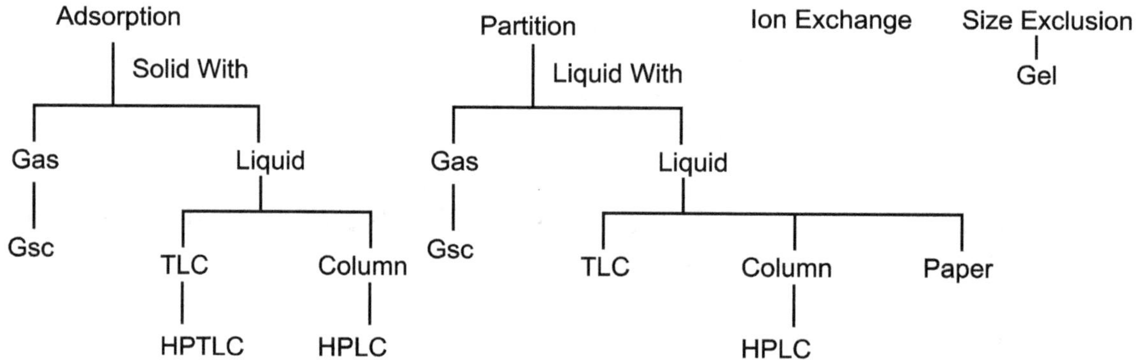

Fig.5.1 Classification Based on Principle of Separation

General techniques involved in Chromatography

PlanarChromatography	ColumnChromatography
Selection of stationary phase	Selection of stationary phase
Selection of Mobile phase	Selection of Mobile phase
Preparation of plate or paper	Preparation of column
Activation of plate	Activation of stationary phase before column packing

Contd...

PlanarChromatography	ColumnChromatography
Application of sample	Application of sample
Chamber saturation	Optimisation of flow rate and fraction volume
Development of chromatogram	Collection of fraction
Location of spots	Monitoring of fractions by Paper Chromatography, TLC or UV-spectroscopy
Evaluation of chromatogram	Quantitative evaluation of fractions

5.3.1.1 Planar Chromatography

The name itself explains that the support for stationary phase is a planar surface like glass plate, aluminium foil, or Whatman filter paper. This is an unavoidable technique in the isolation of phytoconstituents. Without thin layer chromatography (TLC) or paper chromatography, one cannot determine the exact number of constituents present in an extract or isolated product. The general steps involved in planar chromatography i.e. TLC and Paper chromatography techniques are as follow:

Paper Chromatography

Paper chromatography was invented in 1943 by Martin and Synge while separating constituents of plants. It is simplest form of planar chromatography which uses paper (cellulose) as stationary phase and further follows same techniques as that of TLC to develop and evaluate chromatogram. Its major drawback is corrosive chemicals can not be used for detection of separated constituents from mixture.

Thin Layer Chromatography (TLC)

- **Selection of Stationary Phase** According to the nature of the components to be separated, stationary phase can be polar, non-polar, or less polar. It can be organic (charcoal, cellulose) or inorganic (silica gel, alumina, Kieselguhr, diatomaceous earth, magnesia). According to active sites (functional groups present in stationary phase), the adsorbent can be strongly active (alumina), medium active (silica gel), and less active (magnesia, Kieselguhr).

- **Selection of Mobile Phase** Mobile phase can be selected on the basis of the principle "like dissolves like", i.e., according to the nature of components to be separated. Instead of one solvent, select composition of two or more solvents. Selection is also based on technique of elution like isocratic, fractional, or gradient. Generally, volatile solvents should be selected, and highly polar solvents should be avoided.

- **Preparation of Plate** Plates for TLC can be prepared by pouring, dipping, spraying, or spreading methods. Usually, spreading (by readymade spreaders) is the preferred technique as it results in uniform plates. Pouring technique is mostly useful for laboratory scale where the slurry of adsorbent has to be poured at the centre of plate and then, tilted slowly to all corners so as to form a plate of uniform thickness.

- **Activation of Pre-coated Plates** Manually prepared as well as readymade plates should be activated by placing in an oven at 110°C -120°C for 30 minutes prior to sample application. This removes the absorbed water molecules and frees active sites of adsorbent which should be reacted with the sample to be separated. Thus, better resolution and separation can be achieved. Aluminium sheets should be kept in between two glass plates and placed in the oven at 110°C -120°C for 15 min.

- **Pre-conditioning (Chamber Saturation)** It is to ensure uniform distribution of the eluent vapour throughout the chamber prior to chromatographybecause unsaturated chamber causes high retardation factor (Rf) values and dispersion of spots. Saturation of the chamber by lining with filter paper for 30 min prior to development causes uniform distribution of solvent vapours. Thus, less solvent is required for chromatogram development. Chamber can also be saturated by keeping aside for 15-20 min, so that solvent vapours will form equilibrium in chamber.

- **Application of Sample**Sample can be applied eitherby micropipettes, capillaries, micro capillaries, or syringes. In case of HPTLC, being instrumental, automatic sample applicator is available.

- **Development of Chromatogram** Place sample applied plate in a saturated chamber and allow to develop chromatogram upto 8 cm of plate length. Methods used for development can be ascending, descending, circular, two dimensional, multiple, or preparative.

- **Location or Detectionor Visualization ofSpots** Detection under UV (Ultraviolet) light is the first choice which is a non-destructive method. Spots of fluorescent compounds can be seen at 254 nm (short wavelength) or at 366 nm (long wavelength). Spots of non-fluorescent compounds can be seen on fluorescent stationary phase (e.g., Silica gel GF^{254}). Non-UV absorbing compounds can be observed by dipping the plates in 0.1% iodine solution. When individual component does not respond to UV, derivatization by spraying reagents is necessary.

- **Evaluation of Chromatogram:** It can be qualitative (comparison of Rf value of sample with that of the standard) or quantitative. The evaluation can be done by:

  - Comparison of area, intensity of separated spots with that of standard;
  - UV spectroscopic evaluation; and

- **Densitometer:** Comparative determination of optical density of separated bands and blank area where results are obtained in the form of graph i.e. area or height versus Retention factor (Rf) value.

High PerformanceThin Layer Chromatography (HPTLC)

HPTLC is a semiautomatic instrumental TLC. HPTLC is different from conventional TLC due to smaller particles (≤10 μm) of adsorbent and less thickness of the applied layer (≤150 μm). But the striking differences are; it has an automatic sample applicator which is able to apply sample in microlitre amounts; automatic development chamber provided with solvent reservoirs; and densitometer or scanner which scans separated bands by measurement of optical density. Hence, the process is more efficient due to small particle size and no manual errors.

The automatic sample applicators of HPTLC apply sample in the form of bands, and hence, it gives better separation and resolution especially for components having close Rf values like isomers of curcumin. Simple spot application in TLC causes mixing of close Rf value components and chromatogram has a tailing appearance. HPTLC enables simultaneous analysis of many samples in less time with better analytical precision and accuracy.

There is no need of prior treatment like filtration and degassing for solvents in HPTLC. HPTLC plate develops faster in less mobile phase required per sample. There are no chances of contamination from the previous analysis as the fresh stationary and the mobile phase have to be used for each sample. Separated samples can be analyzed visually by simple UV detector. Even non-UV absorbing compounds can be detected by pre or post-chromatographic derivatization.

In case of herbal drugs, it is difficult to isolate and quantify each and every marker component, especially in polyherbal formulations. To overcome this problem, it is necessary to evaluate an herbal product in its entirety which is called "fingerprinting". In this method, HPTLC graphs of each raw material are matched with the HPTLC graphs of the finished product. If the finished product contains maximum peaks of standard material, then it passes the test. Thus, fingerprinting becomes important tool to distinguish between authentic and fake material of herbal drugs.

Comparison of TLC and HPTLC		
Parameter	**TLC**	**HPTLC**
Plates	Handmade/pre-coated	Pre-coated
Adsorbent layer thickness	250um	100-200um
Particle size	5-20 um	4-8 um
Pre-washing plate	Not followed	Must be followed
Application of sample	Manual	Semi-Automatic
Shape	spot	band
Spot size	2-4 mm	0.5-1 mm
Sample volume	1-10 ul	0.2-5 ul
Larger volume	Cannot be applied as it will cause Over loading	Can apply in the form of band
No samples per plate	20	40
Optimum development distance	10-15 cm	5-7 cm
Development time	Depends on volatile nature of mobile phase	50% less than TLC
Reproducibility of results	Difficult	Re-producible

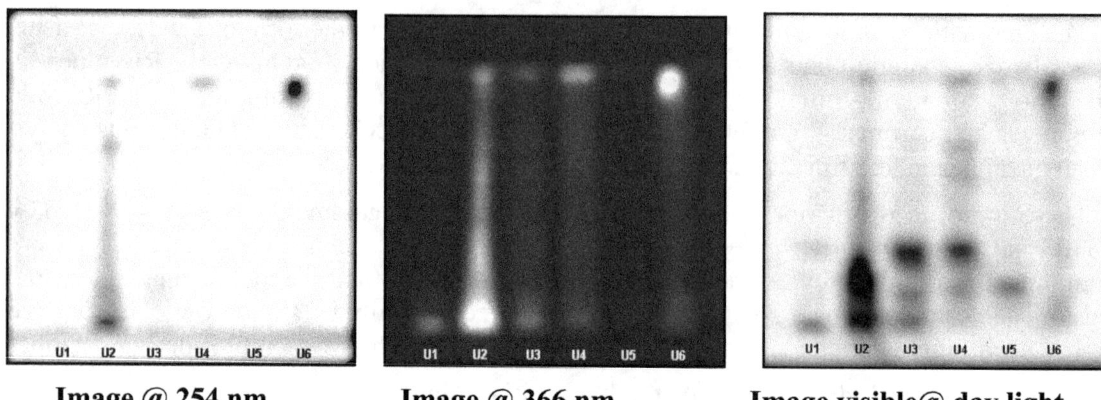

Fig.5.2 Sample HPTLC Photographs by Photo-documentation Device at Different Wavelengths

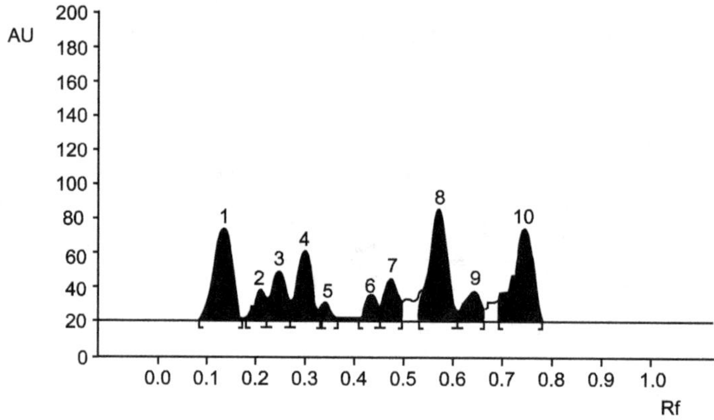

Fig.5.3 Sample HPTLC Graphs of Extracts by Scanning Densitometer

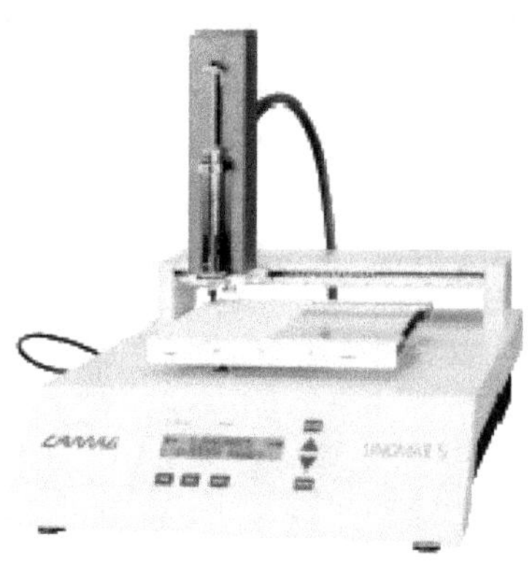

Automatic sample applicator

Automatic development chamber

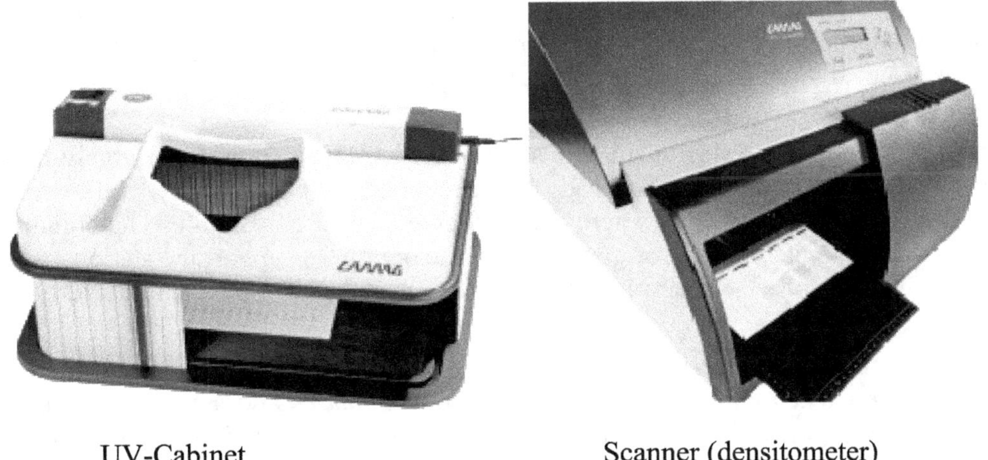

Fig.5.4 Camag HPTLC Instrumentation

5.3.1.2 Column Chromatography

Column chromatography is a method used to purify or isolate phytoconstituents from extracts or partially purified compounds with the help of a column. It is the most preferred method for quantitative or preparative separation. One can isolate the constituents measuring from a few micrograms up to kilograms. In column chromatography, the stationary phase, a solid adsorbent, is placed in a vertical (usually glass) column and the mobile phase, a liquid, is added to the top and flows down through the column either by gravity or external pressure as shown in figure 5.5. The general technique for column chromatography is explained below.

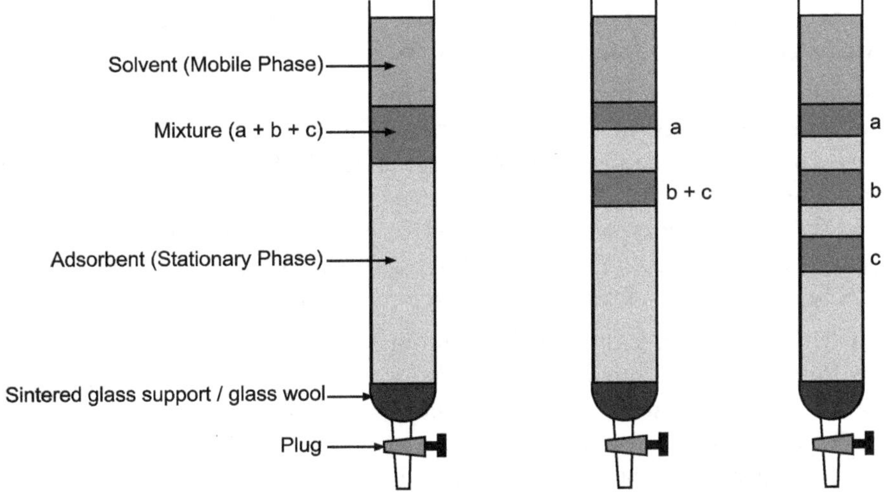

Fig.5.5 Separation by Conventional Column Chromatography

- **Selection of Stationary Phase:** The selection of the stationary phase depends on the polarity of the sample and the mobile phase. For the sample of unknown polarity, the stationary phase is selected through TLC. Generally, 100-500 g of adsorbent is required for per gram of crude extract.

- **Selection of Column:** Select a suitable sized glass or plastic column preferably three times longer than the total volume of the adsorbent required for separation. Metallic columns of packed with adsorbent of 2-20 micrometer particle sizes are available. The column should have the base of sintered glass or glass wool at the bottom to support the stationary phase. In case of reversed phase, separation pre-packed C8 and C18 silica column are the most preferred columns.

- **Column Packing:** Either dry or wet packing are the preferred methods of column packing. However, pre-packed columns of different stationary phases and sizes are available in the market. In wet packing, slurry of the required amount of stationary phase is taken in sufficient amount with least polar solvent from the mobile phase and mixed by stirring until the slurry of good pourability is prepared. In dry packing, the stationary phase is poured in dry form into the column by tapping. Dry packing is particularly useful for VLC of non-polar or of intermediate polarity components where Silica gel 60H is normally used as the adsorbent.In case of gel chromatography, polymers and or polysachharides like agarose, dextran is allowed to swell overnight by suspending the adsorbent in methanol. Porous material with number of different size holes further can be used to separate high larger size particles from small size particles. Generally 1 ml of swollen gel is required for the separation of 1 mg of extract. More non-polar solvents like chloroform or dichloromethane can also be used.

- **Mobile Phase Selection and Addition** Select a proper mobile phase as per nature of constituents. By keeping the outlet open, add the initial mobile phase solvent without disturbing the column. Add enough quantity of solvent, so that adsorbent should be uniformly packed, and the solvent layer should be just above the adsorbent.

- **Sample Application:** For sample application, first put the filter paper or clean dry sand or glass wool over the layer of packed stationary phase to avoid disturbance of column. Dissolve the sample in a small amount of the initial mobile phase solvent and apply carefully to the top of the column bed by using syringe or pipette or capillaries. If the sample is insoluble in the selected solvent or of semisolid nature, then dissolve the sample in a small amount of solvent and mix with silica gel. Evaporate the solvent and load this mass on to the top of the column. Open the outlet valve to allow the sample to get adsorbed into the bed.

- **Elution of Mobile Phase:** Now fill the column with the initial mobile phase solvent and allow it to flow by various methods like nitrogen pressure at the inlet and vacuum at the outlet (e.g., VLC) or pumping the mobile phase through the column at varying pressures (e.g., FC). Gravity elution is a simple method where mobile phase is allowed to run naturally under gravity pressure. Silica gel of 60 mm size should be used for better separation in gravity elution. Positive pressure elution technique involves application of positive pressure to the top of the column to accelerate the flow rate and achieve better resolution. In an alternative method, called vacuum elution, vacuum is applied at the end of the column containing Silica gel 60H. A solvent reservoir is used to maintain the volume of solvent, and the flow rate can be controlled by adjusting the outlet valve. Gradient elution by changing the composition of mobile phase using solvents from the non-polar to polar system gives excellent fractionation of natural compounds. Fill the column with the initial mobile phase required for separation.

- **Collection of Fractions:** Use small glass vials or test tubes or automated fraction collector to collect fractions. Generally, 5-50 ml range fractions are collected. Monitor the fractions by TLC and combine similar TLC pattern fractions. Re-column the fractions to yield purified and isolated constituents.

- **Detectors**: Three types of detectors are commonly used in modern column chromatographic instruments, i.e., Ultraviolet -Visible (UV-Vis), Refractive index, and Photodiode Array (PDA) detectors. A PDA is a linear array of discrete photodiodes on an integrated circuit (IC) chip. For spectroscopy, it is placed at the image plane of a spectrometer to allow a range of wavelengths to be detected simultaneously. In this regard, it can be thought of as an electronic version of photographic film. Array detectors are especially useful for recording the full UV-Vis absorption spectra of samples that are rapidly passing through a sample flow cell, such as in HPLC detector. Diode arrays having a number of elements ranging from 128 to 1024 – and even up to 4096 - are available. This multichannel detector makes an ideal sensor for an entire spectrum in an UV-Vis dispersive spectrophotometer.

High Pressure Liquid Chromatography (HPLC)

HPLC is modified, semi or fully automatic column chromatographic technique (Figure 5.6) based on principle of partitioning. It is different from conventional open chromatographic technique on the basis of following points:

Parameters	Open column	HPLC
Elution method	By gravity at atmospheric pressure which is time consuming	High pressure upto 600 times than atmospheric pressure which helps in fast separation
Particle size adsorbents	40-100 micrometer so separation with less resolution	2-20 micrometer hence separation with high resolution
Length of column	30 cm -150 cm	8 cm -30 cm
Separation time	Long time from hours to days	Very less time even just 10 minutes
Separation sensitivity	Less sensitive	Ultra sensitive
Sample quantity	Milligrams	Nanogram

HPLC allows use of many modified adsorbents which are shown in Fig.5.8 and columns in Fig.5.9 which helps in reverse phase (RP) and normal phase (NP) based separation of polar constituents. Common Stationary Phases and related separation methods useful in HPLC are summarisedas follows.

Table 5.1 Common Stationary Phases and related separation methods useful in HPLC

Separation method	Adsorbent of HPLC
Reversed phase	C18, C8
Normal and reversed phase	CN (cyano), Diol
Normal phase	Silica
Strong cation exchange	Benzene sulfonic acid,

In HPLC, adsorbent is of smaller particle size (as small as 3 μm) but smaller packing requires higher pressure by pump. The use of high pressures in a narrow column allows for a more effective separation to be achieved in much lesser time. Normal flow rate is 0.01-10 ml/min and maximum pressure is 20,000 psi or 600 bar. Pumps should be inert to solvents, buffer salts, and solutes and usually made of stainless steel, titanium, resistant minerals, (sapphires and ruby) and Polytetrafluoroethylene (PTFE) (teflon). A common HPLC detector is a UV absorption detector, as most medium and large molecules absorb UV radiation. Detectors that measure fluorescence and refractive index, are also used for special applications. A relatively new development is the combination of HPLC separation with NMR detector. HPLC can be used in both qualitative and quantitative applications, i.e., for both compound identification and quantification. HPLC separation can be performed now mostly in reverse phase and are used very rarely in normal phase. Reverse phase HPLC (RP-HPLC) is ineffective for only a few separation types; it cannot separate inorganic ions (they can be separated by ion exchange chromatography); it cannot separate polysaccharides (they are too hydrophilic for any solid phase adsorption to occur) and polynucleotides (they adsorb irreversibly to the reverse phase packing).

Two types of pump modules are available in HPLC; isocratic pump and gradient pump. Isocratic pump delivers constant mobile phase composition. It is a low cost pump. Whereas, gradient pump delivers variable mobile phase composition and can be used to mix and deliver an isocratic mobile phase or a gradient mobile phase. Binary gradient pump delivers two solvents and quaternary gradient pump delivers four solvents. Figure 5.7 shows HPLC chromatogram.

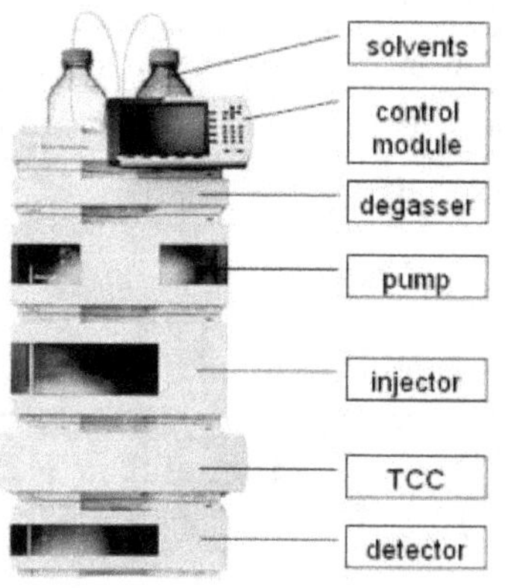

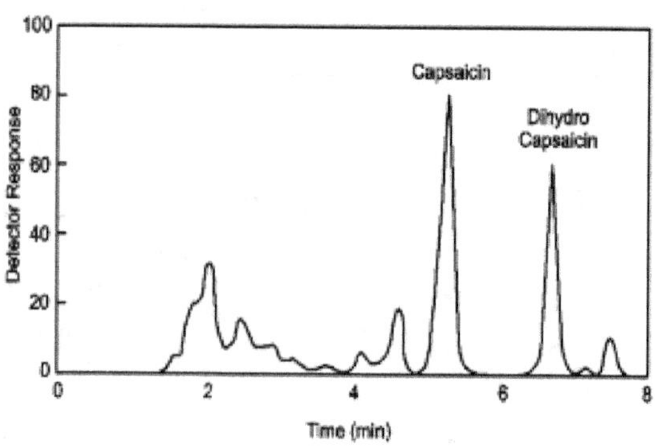

Fig.5.6 Agilent HPLC instrumentation

Fig.5.7 HPLC Chromatogram

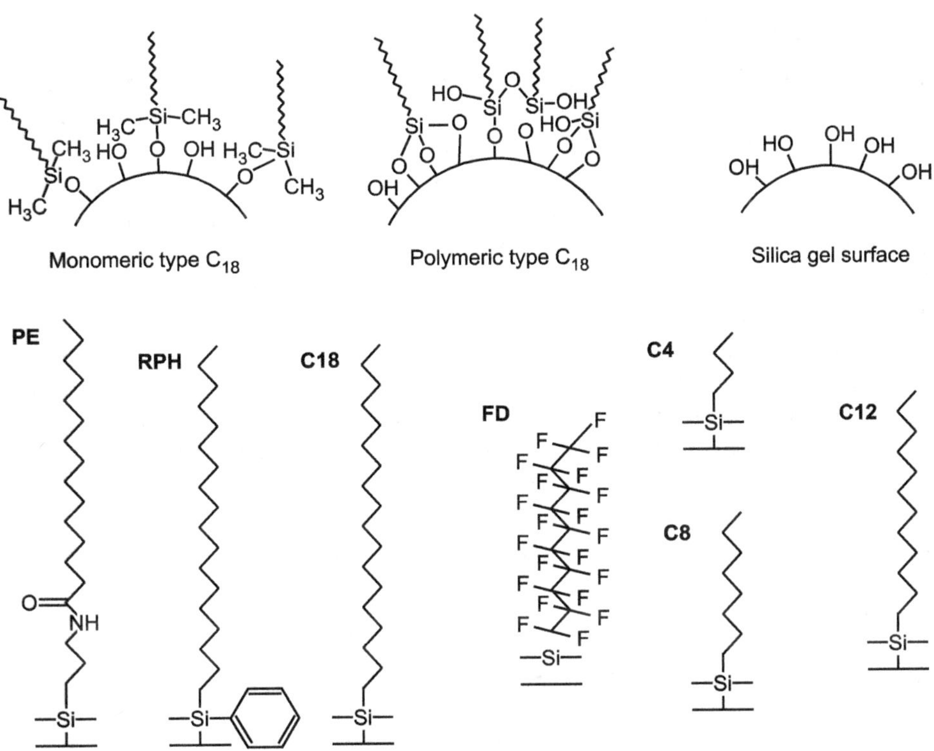

Fig.5.8 Chemistry of modified silica adsorbent

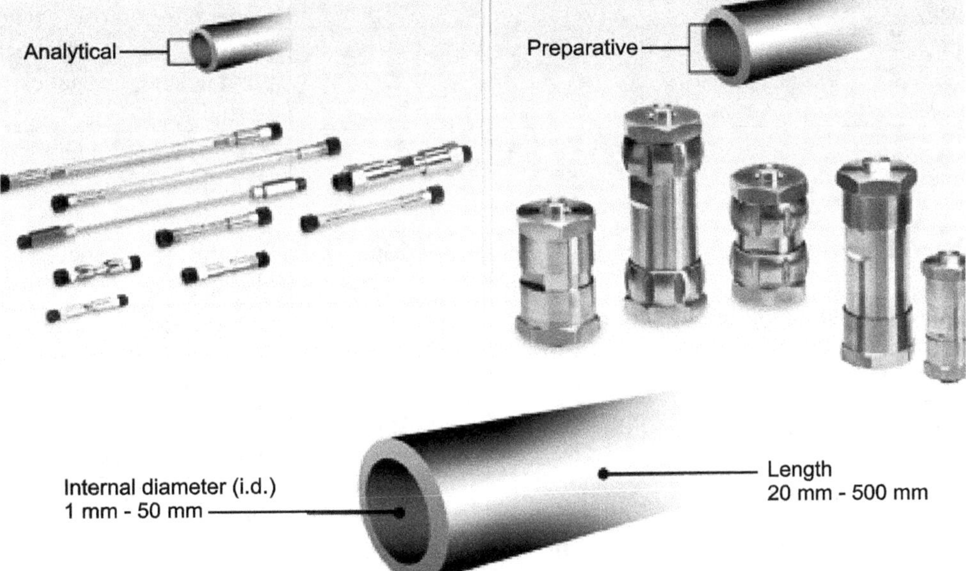

Fig.5.9 Different Types of Readymade Columns

Gas Chromatography (GC)

GC is a common type of column chromatography used in separating and analyzing compounds that can be vaporized without decomposition. In gas chromatography, the mobile phase is a carrier gas, usually an inert gas such as helium or an unreactive gas such as nitrogen, argon, hydrogen, and air. The stationary phase is a microscopic layer of liquid or polymer on an inert solid support inside a piece of glass or metal tubing (known as column) that may be packed or open. Stationary phases are usually bonded and/or cross-linked. Bonding means covalent linking of stationary phase to support, whereas cross-linking means polymerization reactions after bonding with the support. Table 5.2summarised few common stationary phases for Gas – Liquid chromatography (GLC) are summarized below. GC is similar to fractional distillation, since both processes separate the components of a mixture primarily based on boiling point (or vapour pressure) differences. However, fractional distillation is typically used to separate components of a mixture on a large scale, whereas GC can be used on a micro-scale. The most commonly used detectors are flame ionization detector (FID) and thermal conductivity detector (TCD). Since TCD is non-destructive, it can be operated in-series before FID (destructive in nature), thus providing complementary detection of the same analytes. It is very suitable for analysis of essential and fatty oils, especially when attached to mass or NMR spectroscopy. Fig.5.10 shows GC instrument.

Table 5.2 Common Stationary Phases for Gas-Liquid Chromatography

Stationary phase	Common trade name	Maximum temperature, °C	Common applications
Polydimethyl siloxane	OV-1, SE-30	350	General purpose non-planar phase, hydrocarbons, polynuclear aromatics, drugs, steroids
Poly (phynylmethyldimethylsiloxane) (10% phenyl)	OV-3, SE-52	350	Fatty acid methyl esters, alkaloids, drugs, halogenated compounds
Poly (phenylmethylsiloxane)(50% phenyl)	OV-17	250	Steroids, pesticides, and glycols.
Poly (trifluoropropyl-dimethyl) siloxane	OV-210	200	Chlorinated aromatics, nitroaromatics, alkyl-substituted benzenes
Polyethylene glycol	Carbowax 20 M	250	Free acids, alcohols, ethers, essential oils, and glycols
Poly (dicyanoallyldimethyl) siloxane	OV-275	240	Fatty acids, rosin acids, and free acids

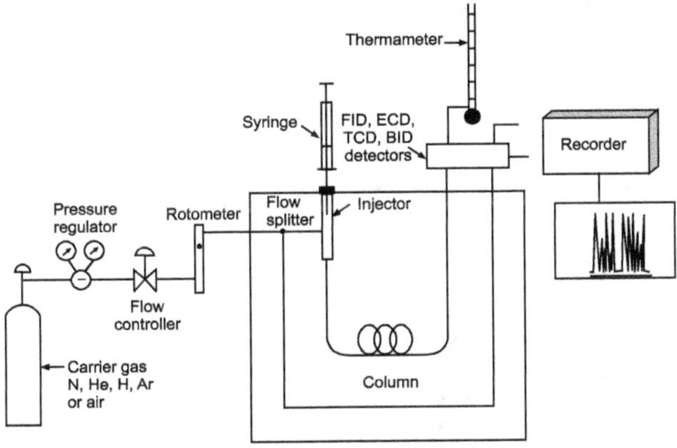

Fig.5.10 Gas Chromatography Instrument outline

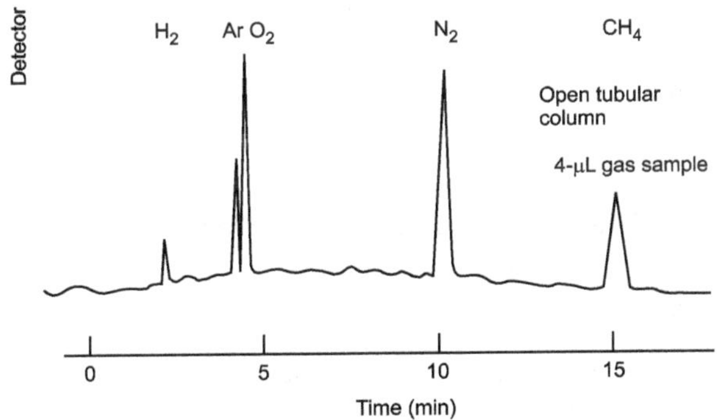

Fig.5.11 GC Chromatogram

Gel Chromatography

Size-exclusion chromatography (SEC) is a chromatographic method in which molecules in solution are separated by their size and molecular weight. It is usually applied to large molecules or macromolecular complexes such as proteins and industrial polymers. Commercial gels like sephadex, bio-gel (cross-linked polyacrylamide), agarose gel, and styragel are often used based on different separation requirements. The most common eluents for polymers that dissolve atroom temperature are tetrahydrofuran (THF), o-Dichlorobenzene, and trichlorobenzene; at 130–150°C for crystalline polyalkenes, m-Cresol, and o-Chlorophcnol; and at 90°C for crystalline condensation polymers such as polyamides and polyesters. The main application of gel-filtration chromatography is the fractionation of proteins, amino acids, enzymes, and other water-soluble polymers. The collected fractions are often examined by detectors. There are many types of detector available and they can be divided into two main categories.

- **Concentration Sensitive Detectors** This includes UV absorption, differential refractometer (DRI) or RI detectors, infrared (IR) absorption, and density detectors.

- **Molecular Weight Sensitive Detectors** This includes low angle light scattering detectors (LALLS) and multi angle light scattering (MALLS).

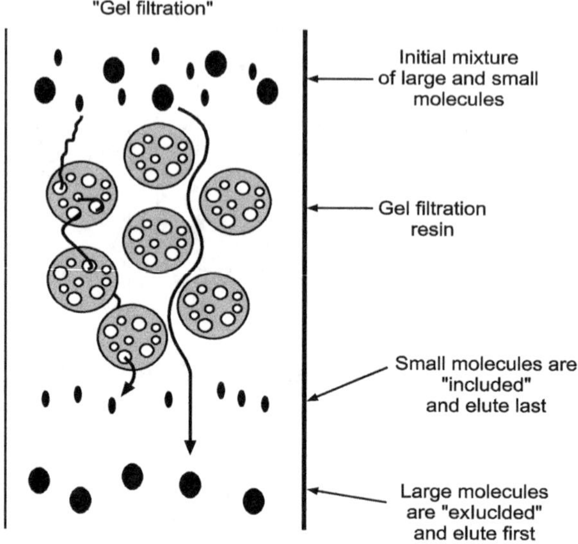

Fig.5.12 Elution by Gel chromatography

Ion Exchange Chromatography (IEC)

IEC is a process that allows the separation of ions and polar molecules based on their charge. It can be used for almost any kind of charged molecule including large proteins, small nucleotides, and amino acids. The solution to be injected is called the sample, and the individually separated components are called analytes. It is often used in protein purification, water analysis, and quality control. IEC retains analyte molecules on the column based on ionic interactions. The stationary phase surface displays ionic functional groups (r-x) that interact with analyte ions of opposite charge. This type of chromatography is further subdivided into cation exchange chromatography and anion exchange chromatography. The analytes of interest must then be detected by their conductivity or UV-Vis light absorbance. Most commercial resins are made of polystyrene sulfonate based on crosslinked polystyrene. Anion resins and cation resins are the two most common resins used in the ion exchange process. While anion resins attract negatively charged ions, cation resins attract positively charged ions. Four main types of ion exchange resins differ in their functional groups:

- Strongly acidic, typically featuring sulfonic acid groups, e.g. sodium polystyrene sulfonate or polyAMPS.

- Strongly basic, typically featuring quaternary amino groups, for example, trimethylammonium groups, e.g.polyAPTAC)

- Weakly acidic, typically featuring carboxylic acid groups.

- Weakly basic, typically featuring primary, secondary, and/or ternary amino groups, e.g. polyethylene amine.

Commonly used ion exchangers in Ion exchange Chromatography			
Sr. No	Name	Type	Functional group
1	DEAE Cellulose (Anion exchanger)	Weakly basic	DEAE (Diethylaminoethyl)
2	QAE Sephadex (Anion exchanger)	Strongly basic	QAE (Quaternary aminoethyl)
3	Q Sepharose (Anion exchanger)	Strongly basic	Q (Quaternary ammonium)
4	CM- Cellulose (Cation exchanger)	Weakly acidic	CM (Carboxymethyl)
5	SP Sepharose (Cation exchanger)	Strongly acidic	SP (Sulfopropyl)
6	SOURCE S (Cation exchanger)	Strongly acidic	S (Methyl sulfate)

Three ion-exchange resins, sodium polystyrene sulfonate, colestipol, and cholestyramine, are used as active ingredients. Sodium polystyrene sulfonate is a strongly acidic ion-exchange resin and is used to treat hyperkalemia. Colestipol is a weakly basic ion-exchange resin and is used to treat hypercholesterolemia. Cholestyramine is a strongly basic ion-exchange resin and is also used to treat hypercholesterolemia. Colestipol and cholestyramine are known as bile acid sequestrants.

Affinity Chromatography

Affinity chromatography, type of column chromatography, is the most efficient separation technique specially developed for those substances which have specific affinity or interactive property with other molecules, e.g., antigen and antibody, enzyme and substrate. This is the only chromatography which uses the biological properties of the substances for separation. The very simple form of affinity chromatography is the classic immune-affinity chromatography related to antibody affinity.

Principle Selective ligands on beaded and porous matrices are used for binding the target compounds. The interactions between ligand and target molecules can be a result of electrostatic or hydrophobic interactions, van der Waals' forces and/or hydrogen bonding.

The target compounds are then recovered under modified conditions by either using a competitive ligand or changing the pH, ionic strength or polarity of the mobile phase.

The requirements for successful affinity chromatography are:

- **Matrix** Chemically and physically inert support for ligand attachment. The common matrices used in affinity chromatography are summarized below.
- **Spacer Arm** For small ligands (MW < 5000), there is a risk of steric hindrance between the ligand and the matrix which restricts the binding of target molecules; hence, pre-activated matrix with spacer arms are used to improve binding between ligand and target molecules. Spacer arm is not necessary for ligands with MW > 5000. Common spacer arms useful in affinity chromatography are 4-12 atom moieties of amino, carboxyl, thiol or hydroxyl groups.
- **Ligand** Molecule that binds reversibly to a target molecule(s). The appropriate ligand is first coupled to the suitable solid support (matrix) and the solution of biologically active products is passed through the column. The compounds having affinity will be retained or adsorbed

by the matrix. Other compounds will pass through the column. The common ligands useful in affinity chromatography are shown in table 5.3.

Table 5.3 Common Matrices and ligands used in Affinity Chromatography

Chemical classes	Examples of Matrices
Polysaccharides	Agarose, dextran, cellulose
Synthetic co-polymers	Polyacrylamide, polystyrene
Inorganic	Iron oxide, porous glass, silica
Biopolymers	Agarose, polysaccharides
Chemical classes	**Examples of ligands**
Enzyme	Enzymes, co-factors, antibodies, substrates analogues
Peptides	Enzymes and antibodies
Proteins	Hormones, vitamins, steroids, lectins, lipids, antibodies, mono-polysaccharides
Nucleic acids	Nucleic acid bases, nucleosides, oligonucleotides
Cells, viruses	Antigens, antibodies, lectins
Lectins	Polysaccharides, hormones, glycoproteins

Common elution buffer systems for protein affinity purification

These conditions apply primarily to protein-protein binding interactions, such as between an antibody and its peptide antigen. Elution buffers for binding interactions between other kinds of molecules may be quite different.

Condition	Buffer
pH	100 mM glycine•HCl, pH 2.5-3.0 100 mM citric acid, pH 3.0 50–100 mM triethylamine or triethanolamine, pH 11.5 150 mM ammonium hydroxide, pH 10.5
Ionic strength and/or Chaotropic effects	3.5–4.0 M magnesium chloride, pH 7.0 in 10mM Tris 5 M lithium chloride in 10mM phosphate buffer, pH 7.2 2.5 M sodium iodide, pH 7.5 0.2-3.0 M sodium thiocyanate
Denaturing	2–6 M guanidine•HCl 2–8 M urea 1% deoxycholate 1% SDS
Organic	10% dioxane 50% ethylene glycol, pH 8-11.5 (also chaotropic)
Specific competitor	>0.1 M counter ligand or analog (Example: using glutathione to elute GST-tagged proteins from immobilized glutathione agarose resin)

Affinity chromatography is a useful device with wide applications in industry, clinical studies, analysis, and biotechnology. It is used to purify constituents like proteins, amino acids, nucleic acids, and sugars from mixtures. In analysis, it is used to study the quantitative interaction and to calculate the dissociation constant. In biotechnology, it is used in the purification of recombinant proteins. In fermentation processes, it is used as process monitor.

Immobilized Metal Affinity Chromatography (IMAC) uses interactions between protein and chelated metal for separation. Recent development in affinity chromatography, i.e., molecular imprinted polymer (MIP) methodology which develops the complementary binding sites (imprints) from known active constituents on selected support. These imprints are further useful to analyze or separate structurally related compounds. The MIP method has the ability to produce binding columns in the absence of known targets.

Flash Chromatography

Flash chromatography or medium pressure chromatography is a fast and efficient separation technique for natural products. It differs from conventional chromatography due to smaller silica gel particles (250-400 mesh) and use of pressurized gas (10-15 psi) to enable rapid and better resolution in separation of phytoconstituents. Pre-packed plastic column (normal or reversed-phase columns) of silica gel and an isocratic or gradient solvent system, which runs with the help of air or pump pressure, enables the mixture to get separated efficiently. The mixtures which show two-three spots in TLC and have Rf values lying between 0.15-0.20 can be separated by flash chromatography. The higher Rf values need higher quantity of solvent. Commonly preferred solvents in flash chromatography are ethyl acetate, dichloromethane, hexane, ether, and methanol. Different flash chromatography instruments are available in the market. Figure 5.13 shows the actual flash chromatography instrument by Teledyne Isco. Following are the characteristics of flash chromatography:

- Fully automated method with user-friendly software;
- Online UV-Vis detection of separation carried out on the packed column;
- Sequential separation;
- Multi-sample parallel purification;
- Purification of organic compounds; and
- Useful in research of agrochemicals, petrochemicals, natural products, polymers, and catalysts.

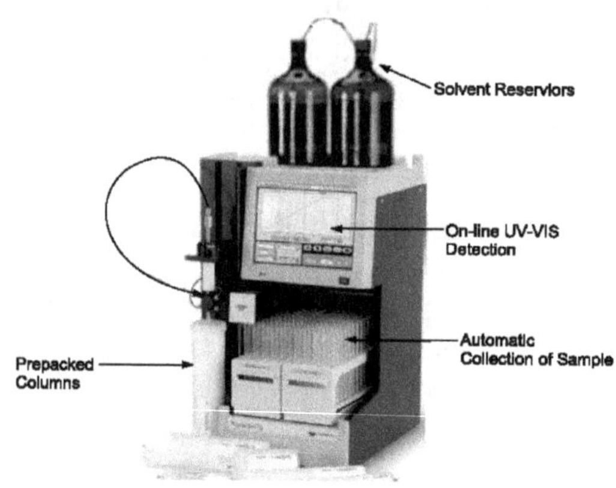

Fig.5.13 Flash Chromatography of *Teledyne Isco*

HyphenatedTechniques

The coupling of a separation technique and an online spectroscopic detection technology is known as hyphenated technique. For example, GC-MS, LC-MS, liquid chromatography-photodiode array (LC-PDA), liquid chromatography-fourier transform infrared spectroscopy (LC-FTIR), liquid chromatography-nuclear magnetic resonance spectroscopy (LC-NMR), and liquid chromatography-nuclear magnetic resonance spectroscopy-mass spectroscopy (LC-NMR-MS). These methods are useful for quality control, analysis, and chemotaxonomic and metabolic studies of phytochemicals. Sometimes, more than two techniques are coupled for efficient separation like LC-PDA-MS, PDA-NMR-MS, and SPE-LC-MS. The coupling of HPLC and SPE, knwon as sepbox, is one of the most efficient and sophisticated hyphenated techniques for fractionation and isolation of phytoconstituents.

Applications of Chromatography				
Technique	**Stationary phase**	**Mobile Phase**	**Detector**	**Application**
Planar chromatographic techniques				
Paper	Whatman paper (Cellulose)	Organic solvents	UV-cabinet Or any non-corrosive derivitisation spraying agents	Identification Separation Purification Impurity detection (Qualitative and preparative)
TLC	Silica, alumina, calcium carbonate, starch,	Organic solvents	UV-cabinet Or any derivitisation spraying agents	Identification Separation Purification Impurity detection **(Qualitative and preparative)**

Contd...

HPTLC	Silica, alumina, calcium carbonate, starch,	Organic solvents	Densitometer	**Qualitative and quantitative:** Identification Separation Purification Impurity detection **(Fingerprint)**
Columnchromatographic techniques				
HPLC	Normal phase And reversed phase **Silica (C8, C18, Cyano, Diol)**	Organic solvents	UV-Detector, Photo-diode array detector, Refractive index detector	Identification Separation Purification Impurity detection Isolation of Phytochemicals
Gas	Polyethylene glycol, Polydimethyl siloxane, Poly (phynylmethyldimethylsiloxane) (10-50% phenyl) In gas liquid chromatography (GLC) generally non-volatile liquid is to be coated as a thin film on solid support like Diatomaceous earth)	Gases like nitrogen, helium, argon, air	Thermal Conductivity detector (TCD) Flame Ionization detector (FID) Catalytic combustion detector (CCD) Electron capture detector (ECD) Mass spectrometer (MS) Infrared detector (IRD)	Identification Separation Purification Impurity detection **(Gaseous and low molecular weight natural chemicals like volatile as well as non-volatile oils)**
Gel	Gelling materials like dextran, agarose	Organic solvents	UV-Detector	Identification Separation Purification Impurity detection **(Isolation of High molecular weight phytochemicals like proteins, polymers, enzymes**
Ion exchange	large quantity of acid/basic groups attached to a	Inorganic salt dissolve	Conductivity	Softening and demineralisation of water.

Contd...

| | polymeric resin | d in a suitable solvent, | | | ◘ For extraction of enzymes from tissues.
◘purification of solutions free from ionic impurities.
◘separation of inorganic ions.
◘ separation of sugars, amino acids and proteins |
| **Affinity** | Sepharose, Agarose, Cellulose | Buffers | Electrophoresis | | It is used to purify constituents like proteins, amino acids, nucleic acids, and sugars from mixtures. |

5.3.2 Electrophoresis

Electrophoresis is an easy and comparatively inexpensive, and powerful molecular separation technique. It can be defined as the movement of charged particles or colloidal aggregates relative to the electrolyte under the influence of an electric field. The rate of this migration depends on properties like net charge of ions, molecular weight, size, temperature, solubility, viscosity, properties of support medium, ionic strength of the buffer, and magnitude of current applied.

Principle Electrophoresis technique is based on the principle that charged particles in a liquid media under the influence of an electric field will migrate to the electrode of the opposite charge. Positive ions (cations) will migrate to the cathode (negative electrode) and negative ions (anions) will migrate to the anode (positive electrode). The molecules get the force of the applied electric field pulling them in one direction which is given by

$$F = EQ$$

Where, F is the force,

E is the electric field,

and   Q is the charge on the molecule.

Thus, greater the charge on the molecule, the greater the force requires and molecules will move faster. However, molecules with a larger mass will move slowly. It clears from the fact that the electrophoretic mobility of a molecule depends on its charge to mass ratio.

Requirements

Electrophoresis requires support medium like whatman filter paper 1 or 3, agar gel, agarose, cellulose, or polyacrylamide gel. Buffers are required for two important reasons; one is to provide the pH which imparts charge on the solute to be separated; and second is to conduct the separation under influence of current. The pH ionizes solutes and the resulting net charge determines the flow of migration of solute to any one of the electrodes. Buffers like barbital and tris-borate-EDTA are commonly used, which fix the effective pH at 8.6 leading to sharper bands and good separations. In addition to the above, power supply, positive and negative electrodes, chamber, and identification or detection method are needed for electrophoresis. Table 5.4 explains various chemicals and their role in electrophoresis.

Table 5.4 Chemicals and their Role in Electrophoresis

Chemical	Role
Polyacrylamide gel (PAG)	Synthetic gel, versatile medium separates protein molecules according to both size and charge.
Agarose	Naturally occurring linear polysaccharide; particularly useful for the separation of large sequences of DNA
Acrylamide	Crystalline powder, viscous but does not form a gel
N, N, N', N'-tetramethylethylenediamine(TEMED)	Chemical polymerisation of acrylamide gel
Bisacrylamide (N,N'-Methylenebisacrylamide)	Cross linking agent for polyacrylamide gels
Ammonium persulphate (APS)	Initiator for gel formation
Sodium dodecyl sulphate (SDS)	Most common dissociating agent used to denature native proteins to individual polypeptides. All the proteins become negatively charged by their attachment to the SDS anions.
Bromophenol blue (BPB)	Visualising agent, universal marker dye
Coomassie brilliant blue (CBB)	Visualising agent, popular protein stain
n-Butanol	Overlay solution on the resolving gel
Dithiothreitol (DTT)	Reducing agent used to disrupt disulphide bonds to ensure the protein is fully denatured before loading on the gel
2-mercaptoethanol (beta-mercaptoethanol/BME),	Reducing agent
Glycerol	Preservative and a weighing agent
THAM [tris (hydroxy methyl) aminomethane]	Satisfactory buffer with pH ranges from 7-9
Glycine (Amino acetic acid)	Source of trailing ion or slow ion
Silver staining	Uses silver colloid to locate and determine quantity of proteins; this staining increases the sensitivity 50 times than other staining procedures

Types of Electrophoresis

Although the complete description of types of electrophoresis is out of the scope of the book, the gistof some of them are given below.

- **Free Solution Electrophoresis** It is also called as moving boundary electrophoresis where separation is done without any support and measurement is dependent on sharply defined boundaries.

- **Zone Electrophoresis** It isalso called as electrochromatography where electrophoretic separation is done by using supports like paper or gel and fractions separate in discrete forms (zones).

- **Vertical Electrophoresis** It consists of a V-shaped apparatus called vertical model of Darrum. This model requires less amount of buffer and less time for separation. The mechanism and the apparatus for vertical electrophoresis is shown in figure 5.14.

- **Affinity Electrophoresis** It is the method of choice to obtain both qualitative and quantitative information. Here, the interaction or binding of a molecule with the substance having specific affinity will normally change the electrophoretic properties of a molecule.

- **One Dimensional Electrophoresis (1-D)** It is used for most routine protein and nucleic acid separations.

- **TwoDimensional Electrophoresis (2-D)** 2-D electrophoresis begins with 1-D electrophoresis, but then, separates the molecules by a second property in a direction of 90° from the first. It is based on the principle that it is impossible that two molecules will be similar in two distinct properties. Hence, the 2-D causes resolution based on two properties; one is always charge; and other may be isoelectric point or mass. This technique is useful in finger printing and resolving all of the proteins present within a cell.

- **Column Electrophoresis** Any electrophoresis which uses the column to support gel material called as column electrophoresis. This method is less prone to lateral movement of separating components and thus, causes slightly improved resolution of the bands.

- **Slab Gel Electrophoresis** It uses flat gels formed between two plates of a glass. It allows 2-D analysis and of running multiple samples simultaneously in the same gel. This technique is of choice for any blot and or autoradiographic analysis.

- **Discontinuous Gel Electrophoresis** It is also called as disk electrophoresis. It uses multiple gel systems where separating gel is augmented with a stacking gel and an optional sample gel. These gels can have different concentrations of the same support media or may be completely different agents.

- **Gel Electrophoresis** When supporting medium is gel like agar, agarose, polyacrylamide then it is called as gel electrophoresis.

- **Paper Electrophoresis** Here supporting medium is paper like Whatman filter paper 1 and 3 composed of cellulose or cellulose acetate.

- **Sodium Dodecyl Sulfate (SDS) Electrophoresis** SDS is a protein denaturing detergent that causes the molecule to unfold with a uniform negative charge which enables the separation of proteins by their size. Without SDS, different proteins with similar molecular weights would migrate differently due to differences in mass by charge ratio, as each protein has an isoelectric point and molecular weight particular to its primary structure.

- **Sodium Dodecyl Sulphate-Polyacrylamide Gel Electrophoresis (SDS-PAGE)** SDS-PAGE has become a standard means for molecular weight determination of proteins. In this electrophoresis, samples have identical charge per unit mass as binding of SDS results in fractionation by size.

- **Reducing SDS-PAGE** With the addition of SDS, proteins are again treated with a reducing agent such asDithiothreitol (DTT) or Beta-mercaptoethanol (BME), which further reduces disulfide linkages and unfolds tertiary structures of proteins. This method is useful in separating native metalloproteins in complex biological matrices, e.g., quantitative preparative native continuous polyacrylamide gel electrophoresis (QPNC-PAGE),

- **Native Gel Electrophoresis** Native gels are similar to gel electrophoresis except that the detergent (SDS) is not used to denature proteins. Proteins remain in their native state and are separated in the electric field following their mass and the mass of their complexes respectively. Native gels are only able to separate proteins up to 2000 kDa in size.

- **Capillary Electrophoresis (CE)** It is also known as capillary zone electrophoresis (CZE) which can be used to separate ionic species based on their size to charge ratio in the interior of a small capillary filled with an electrolyte. This technique combines electrophoresis and HPLC. The separation of compounds by capillary electrophoresis is dependent on the differential migration of analytes in an applied electric field, i.e., electrophoretic mobilities. CEcan be directly coupled with mass spectrometers or surface enhanced raman spectroscopy (SERS).

- **Horizontal Electrophoresis** This consists of a horizontal box which is divided into two compartments by a platform in the middle. The disadvantage of this system is that it is not possible to use discontinue buffer systems, and as the gels are not covered, atmospheric oxygen does not allow acrylamide to polymerize in this system. Thus, this is suitable for agarose gel only. Figure 5.15 shows the mechanism and apparatus for horizontal electrophoresis.

General Procedure for Gel Electrophoresis

- The fragments obtained after cutting with restriction enzymes are separated by using gel electrophoresis.

- Electric field is applied to the electrophoresis matrix (commonly agarose gel) and negatively charged DNA fragments move towards the anode.

- Fragments separate according to their size by the sieving properties of agarose gel. Smaller the fragment, farther it moves.

- Staining dyes such as ethidium bromide followed by exposure to UV radiations are used to visualize the DNA fragments.

- DNA fragments are visible as bright orange coloured bands in the agarose matrix.

- These bands are cut from the agarose gel and extracted from the gel piece (elution).

- DNA fragments are purified and these purified DNA fragments are used in constructing recombinant DNAs

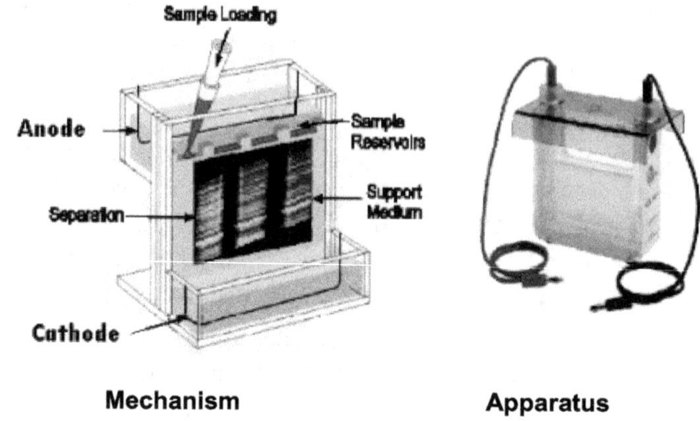

Fig.5.14 Vertical Electrophoresis

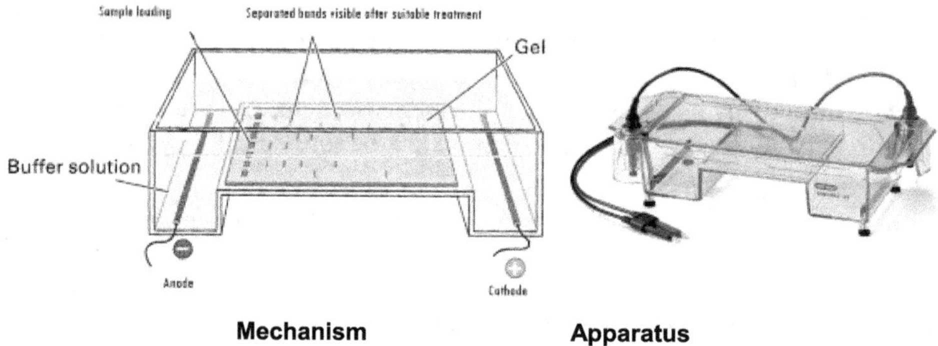

Fig.5.15 Horizontal Electrophoresis

Application: Electrophoresis technique is useful in separation of electrically charged ions like amino acids, proteins, nucleic acid, enzymes, inorganic components, isoenzymes, immunoglobulins, and many other high molecular weight compounds.

5.3.3 Spectroscopy

Spectroscopy can be defined as an analytical technique that uses interaction between the matter and electromagnetic radiations to perform analysis. The instrument which measures this interaction is called a spectrophotometer. Electromagnetic radiation, e.g., gamma rays, X-rays, UV, visible, infrared (IR), microwave, and radio is a form of radiant energy which exhibits both wave (reflection, refraction, and interference) and particle properties (definite energies).Spectrophotometer is a device which measures %transmittance of light radiation when light of certain intensity and of specific frequency range is passed through the sample. Thus, the instrument compares the intensity of transmitted light with that of incident light. This spectroscopy can be absorption or emission spectroscopy.

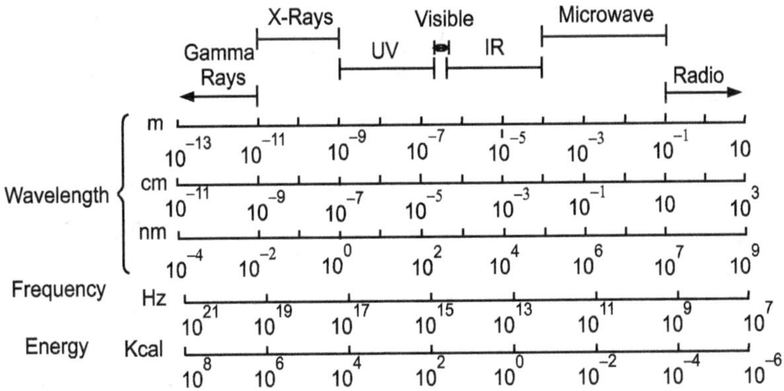

Fig.5.16 Electromagnetic Spectrum

Fundamental Laws of spectrophotometry: The following are the fundamental laws of **spectrophotometry:**

- **Beer's law** This law states that absorbance of the incident light is directly proportional to the concentration of a solution.

- **Lambert's/Bouger'slaw**This law states that absorbance of the incident light is directly proportional to the length of the light path (thickness of the cell).

- As 'e' and 'b'given in the formula are constant, hence,absorbance (A) is directly proportional to concentration (c).

$$A = log_{10}\frac{I_0}{I} = e * b * c$$

Where, A=absorbance,I_0=intensity of incident light,I =intensity of transmitted light,e = molar absorptivity or molar extinction coefficient, b = thickness of cell, c = concentration

General Principle

Molecular electronic transitions take place when valence electrons in a molecule are excited from one energy level to a higher energy level. The energy change associated with this transition provides information on the structure of a molecule and determines many molecular properties such as colour. The following molecular electronic transitions exist:

$$\sigma \rightarrow \sigma^* \; ; \pi \rightarrow \pi^* \; ; n \rightarrow \sigma^*; \; n \rightarrow \pi^*; \; \text{aromatic } \pi \rightarrow \text{aromatic } \pi^*$$

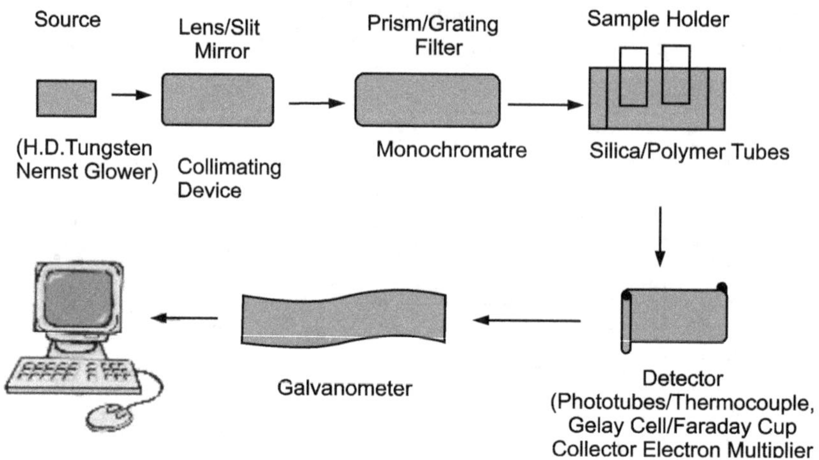

Fig.5.17 Block Diagram for Spectrophotometer

In spectroscopy, the radiations of a known intensity are made to fall on the sample. These radiations transfer their energy to valence electrons of the sample. Valence electrons jump to an excited state and come back to the ground state. During this transition, absorption and emission of energy provided by incident radiation occurs which is characteristic of that sample molecules. The detector receives (Fig.5.16) these transmitted radiations with altered intensity. The change in intensity along with other parameters is employed to produce a spectrum which determines the physical and structural specifications of the sample.

Types/Classification

➢ **Absorption spectroscopy**: Absorption occurs when energy from the radiative source is absorbed by the material. Example: UV- Visible (UV-Vis absorption spectroscopy), IR (Infrared absorption spectroscopy), MASS Spectroscopy

➢ **Emission spectroscopy**: Emission indicates that radiative energy is released by the material. A material's blackbody spectrum is a spontaneous emission spectrum determined by its temperature. *Example*: infrared, flame or fluorescence

➢ **Elastic scattering and reflection spectroscopy** determine how incident radiation is reflected or scattered by a material. *Example:* Crystallography employs the scattering of high energy radiation, such as x-rays and electrons, to examine the arrangement of atoms in proteins and solid crystals.

➢ **Inelastic scattering** phenomena involve an exchange of energy between the radiation and the matter that shifts the wavelength of the scattered radiation. *Example*: Raman and Compton scattering.

➢ **Coherent or resonance spectroscopy** aretechniques where the radiative energy couples two quantum states of the material in a coherent interaction that is sustained by the radiating field. The coherence can be disrupted by other interactions, such as particle collisions and energy transfer, and so often require high intensity radiation to be sustained. *Example*: Nuclear magnetic resonance (NMR) spectroscopy is a widely used resonance method

Typical Instrumentation

Sepectscopy Technique	Radiation source	Detectors
UV-Visible	Tungsten filament (300–2500 nm), a deuterium arc lamp, which is continuous over the ultraviolet region (190–400 nm), Xenon arc lamp, which is continuous from 160 to 2,000 nm; or more recently, light emitting diodes (LED) for the visible wavelengths.	Photomultiplier tube, linear photodiode array, Charge-Coupled Devices (CCDs)
IR/FTIR	Silicon Carbide	Pyroelectric detectors
Raman	Laser	Germanium or Indium gallium arsenide (InGaAs)

Ultraviolet (UV) and Visible Spectroscopy

Ultraviolet -Visible Spectroscopy is concerned with the study of the absorption of UV-visible radiation. Radiation wavelength ranges from UV radiation: 200-400 nm, visible light: 400-800 nm. Thus, for colorless compounds, measurements are made in the range 200-400 nm and for colored compounds, the range is 400-800 nm.

Principle: In both UV as well as visible spectroscopy, only valence electrons absorb the energy which causes promotion of electrons of the sample molecule from the ground state to higher energy states. This is called transition due to which characteristic spectrum is obtained according to nature of substance.The absorption is characteristic and depends on the nature of electrons present. The intensity depends on the concentration and path length.

MajorApplications: The major applications of this spectroscopy are identification of plant constituents, spectral elucidation of organic compounds, and determination of impurities.

Qualitative Analysis: The qualitative analysis is done by measuring λ_{max}.For a given substance, the wavelength at which maximum absorbance in the spectrum occurs is called λ_{max}, pronounced "Lambda-max".

Quantitative Evaluation: For the quantitative evaluation of a substance, a standard calibration curve is first prepared by measuring the optical densities of a series of standard solutions of the pure compound by use of light of asuitablewave length; usually the compound gives an absorption maximum.Graph of UV-visible spectroscopy is of two types, i.e.,wavelength vs. absorbance to obtain λ_{max}; and concentration vs. absorbance to find out unknown concentration of constituents.

Calculation of λ_{max}

- If the parent molecule is heteroannular, start with 214.
- If the parent molecule is homoannular containing two conjugated double bond in six member ring, begin with 253.
- Add 30 for each double bond.
- If system is in *cis* position then add 39 nm. If *trans,* addition is nil.
- Add 5 for each carbon substituent on the double bonds.
- Add 5 for exocyclic double bonds.
- If conjugated carbonyl system is present, then start with 215.
- If *Alpha* organic substituent, add 10; In case of *Beta,* add 12; and if *Gamma/delta,* add 18.
- If exocyclic double bond present, then add 5.
- If homodiene system, then add 39.
- For halogen substitutions, add 5.
- The final number you will obtain is maximum absorbance wavelength (lambda max) in nm.
- If system is benzoic derivative with alkyl substitution, then base value is 246.
- If system is benzoic derivative with hydroxyl substitution, then base value is 230.

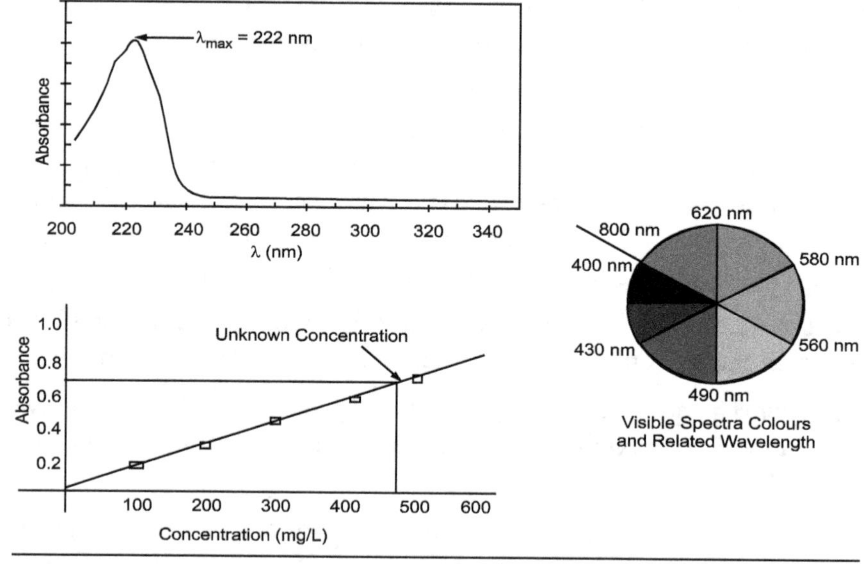

Fig. 5.18 UV-Visible spectrum showing λ_{max} value, calibration curve to know unknown concentration and wavelength ranges for different colors

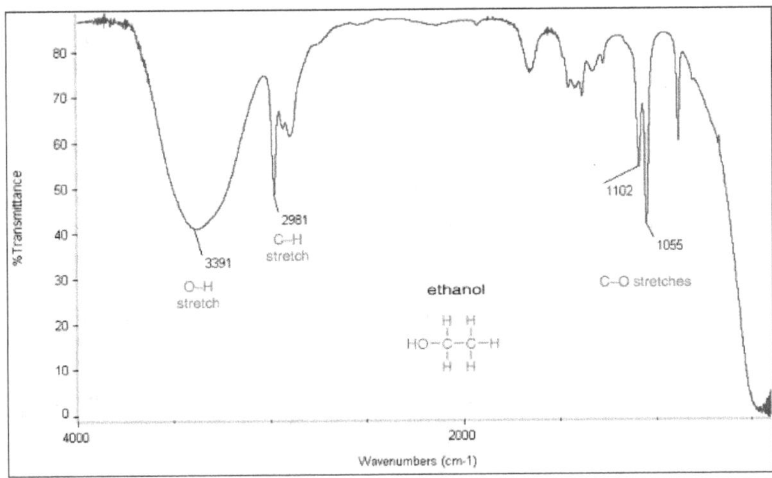

Fig.5.19 IR Spectrum of ethanol

Infrared (IR) Spectroscopy

The infrared portion of the electromagnetic spectrum ($700–14000$ cm^{-1}) is divided into three regions; the near-, mid-, and far-infrared, named for their relation to the visible spectrum.The far-infrared, approximately $400\text{-}10$ cm^{-1} ($1000–30$ μm), lying adjacent to the microwave region, has low energy and may be used for rotational spectroscopy. The mid-infrared, approximately $4000\text{-}400$ cm^{-1} ($30–1.4$ μm), may be used to study the fundamental vibrations and associated rotational-vibrational structure. The higher energy near-infrared, approximately $14000\text{-}4000$ cm^{-1} ($1.4–0.8$ μm), can excite overtone or harmonic vibrations.

   Principle: Infrared spectroscopy exploits the fact that molecules have specific frequencies at which they rotate or vibrate corresponding to discrete energy levels (vibrational modes). Thus, when a molecule gets exposed to IR radiations, it vibrates to higher vibrational/ rotational energy levels and gives characteristic spectrum made of peaks which is due to functional groups present in the molecule. IR spectrum especially in range of $400\text{-}4000$ cm^{-1} is considered as a fingerprint region due to fact that peaks in this region are very specific and characteristic to particular chemical functional group and thus useful in structural elucidation as well as confirmation of identity of chemical structures. .IR spectra involves fundamental as well as overtone vibrations such as stretching (asymmetric, symmetric) and bending (scissoring, rocking, wagging, twisting).

Applications

- Major application is functional group detection;
- Detection and identification of impurity as presence of impurity reduces sharpness of peak or causes extra band;
- Detection of moisture content;
- Shape of the molecule:whether linear or non linear.
- Completion of a chemical reaction.

Table 5.5 IR Spectrum Values for various Functional Groups

Wave Number (cm^{-1})	Functional Group	Associated Groups
690–515	C–Br stretch	Alkyl halide
700–610	–C≡C–H: C–H bend	Alkynes
725–720	C–H rock	Alkanes
850–550	C–Cl stretch	Alkyl halides
900–675	C–H "oop"	Aromatics
910–665	N–H wag	1°, 2° amines
950–910	O–H bend	Carboxylic acids
1000–650	=C–H bend	Alkenes
1250–1020	C–N stretch	Aliphatic amines
1300–1150	C–H wag (–CH 2 X)	Alkyl halides
1320–1000	C–O stretch	Alcohols, carboxylic acids, esters, ethers
1335–1250	C–N stretch	Aromatic amines
1360–1290	N–O symmetric stretch	Nitro compounds
1370–1350	C–H rock	Alkanes
1470–1450	C–H bend	Alkanes
1500–1400	C–C stretch	Aromatics
1550–1475	N–O asymmetric stretch	Nitro compounds
1600–1585	C–C stretch	Aromatics
1650–1580	N–H bend	1° amines
1680–1640	–C=C–	Stretch alkenes
1710–1665	C=O stretch	α, β–unsaturated aldehydes, ketones
1715	C=O stretch	Ketones, saturated aliphatic
1730–1715	C=O stretch	α, β–unsaturated esters
1740–1720	C=O stretch	Aldehydes saturated aliphatic
1750–1735	C=O stretch	Esters saturated aliphatic
1760–1690	C=O stretch	Carboxylic acids
1760–1665	C=O stretch	Carbonyls (general)
2260–2100	–C≡C– stretch	Alkynes
2260–2210	C≡N stretch	Nitriles
2830–2695	H–C=O: C–H stretch	Aldehydes
3000–2850	C–H stretch	Alkanes
3100–3000	=C–H stretch	Alkenes

Contd…

Wave Number (cm^{-1})	Functional Group	Associated Groups
3100–3000	C–H stretch	Aromatics
3330–3270	–C≡C–H: C–H stretch	Alkynes (terminal)
3300–2500	O–H stretch	Carboxylic acids
3400–3250	N–H stretch	1°, 2° amines, amides
3500–3200	O–H stretch	h–bonded alcohols, phenols

NMR Spectroscopy

Nuclear magnetic resonance spectroscopy, abbreviated as NMR spectroscopy, is related to the magnetic properties of certain nuclei. Example includes the most commonly measured nuclei like ^1H and ^{13}C, and rare like ^{113}Cd, ^{15}N, ^{19}F, ^{31}P, ^{17}O, ^{29}Si, ^{11}B, ^{23}Na, ^{35}Cl, ^{195}Pt. Here ^1H-NMR and ^{13}C-NMR has been covered from the point of structural elucidation. NMR is applicable to any nucleus having an odd atomic number or odd mass number because that nucleus has property of nuclear spin.This spin is measured with a nuclear spin quantum number (I) which should always be a positive integral number.

- In case of odd mass number (MN)-odd/even atomic number (AN), I= half integral

- In case of even MN-odd AN, I= 1, 2, 3

- In case of even MN-even AN, I= 0

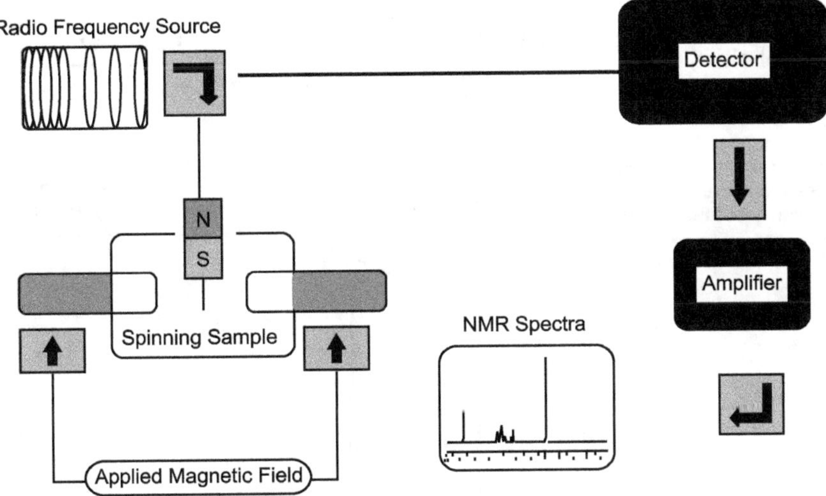

Fig.5.20 Block Diagram of NMR Spectrophotometer

Principle: The nuclei which contain odd numbers of protons or neutrons have an intrinsic magnetic moment and angular momentum due to which they are able to spin. This nuclear spinning generates a magnetic field. When placed in an external magnetic field, spinning protons act like bar magnets, and the magnetic fields of spinning nuclei will align either with external field or against the field; that means about the direction of the magnetic field with an angular frequency (*Larmor frequency*). If precessing nuclei is irradiated with radiofrequency (RF)waves, then through absorption and emission of right RF, energy at the resonance frequency (*Larmor equation*) will cause the spinning proton to flip. This energy difference matched with correct frequency (also known as resonance) of electromagnetic radiation is proportional to strength of the applied magnetic field.

The shift of the NMR frequency due to the chemical environment is called the chemical shift.Circulating electrons present in molecule create secondary induced magnetic field (shielding) that oppose the external magnetic fieldand the nucleus is, therefore, said to be shielded. The chemical environment of protons in a molecule determines the amount of shielding. More shielded absorb at higher field or up field (lower chemical shift) and less shielded (moredeshielded) absorbs at lower field or downfield (higher chemical shift). The number of signals shows how many different kinds of protons are present. Location shows how much shielded ordeshielded.Intensity shows the number of protons of a particular type. Signal splitting shows the number of protons on adjacent atoms. The area under each peak is proportional to the number of protons.

Chemical shift is explained in terms of Delta (Δ) or tau (τ) value (tau =10 minus delta).The chemical shift is reported as a relative measure from some reference resonance frequency. For the nuclei ^{1}H, ^{13}C, and ^{29}Si, TMS (tetramethylsilane) is commonly used as a reference. TMS protons are highly shielded,andhence,the signal defined as zero.This difference between the frequency of the signal and the frequency of the reference is divided by frequency of the reference signal to get the chemical shift.

$$\text{Chemical shift, } \partial = \frac{\text{Frequency of Signal} - \text{Frequency of Reference}}{\text{Spectrometer Frequency}} \times 10^6$$

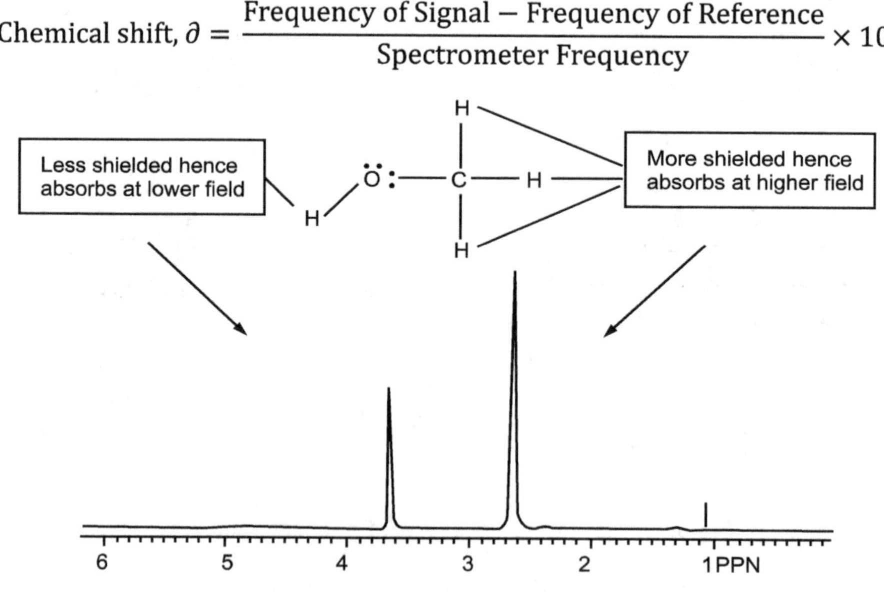

Fig.5.21 Chemical shifts in NMR spectrum

Types of NMR Spectroscopy

Multi-Dimensional NMR spectroscopy

The multi-dimensional NMR spectroscopycan be two, three or four dimensional NMR spectroscopy developed for more reliable interpretation of NMR spectra. Most famous among all is twodimensionalNMR spectroscopy.The conventional, one-dimensional (1D) NMR spectra are plots of intensity vs. only one frequency; but in two-dimensional (2D) spectroscopy, intensity is plotted as a function of two frequencies, commonly called as F1 and F2. The 1DNMR spectra are more complex and most of the peaks overlap each other.Hence, the simpler and elaborative peaks from 2D NMR are easy for accurate structural elucidation. In   IDNMR, the signal is recorded as a function of one time variable, and then transformed to give aspectrum which is a function of one frequency variable. Whereas in 2DNMR, the signal is recorded as a function of two time variables, t1 and t2, and the resulting data transformed twice to yield a spectrum which is a function of two frequency variables.

Techniques Employed in NMR Spectroscopy

Homonuclear: Determination of connectivity between two similar nuclear species, i.e.,^{1}H-^{1}H NMR Following are few homonuclear NMR spectroscopy techniques <ul><li>J-spectroscopy</li><li>Exchange spectroscopy (EXSY)</li><li>Nuclear Overhauser effect spectroscopy(NOESY)</li><li>Total correlation spectroscopy (TOCSY)</li></ul>	**Heteronuclear:**Determination of connectivity between two different nuclear species like proton and carbon or nitrogen and carbon, etc. Following are few heteronuclear NMR spectroscopy techniques <ul><li>Heteronuclear Multiple Quantum Coherence (HMQC)</li><li>Heteronuclear Multiple Bond Coherence (HMBC)</li><li>Heteronuclear Single Quantum Coherence (HSQC)</li><li>Rotating frame Overhause Effect Spectroscopy (ROESY)</li></ul>

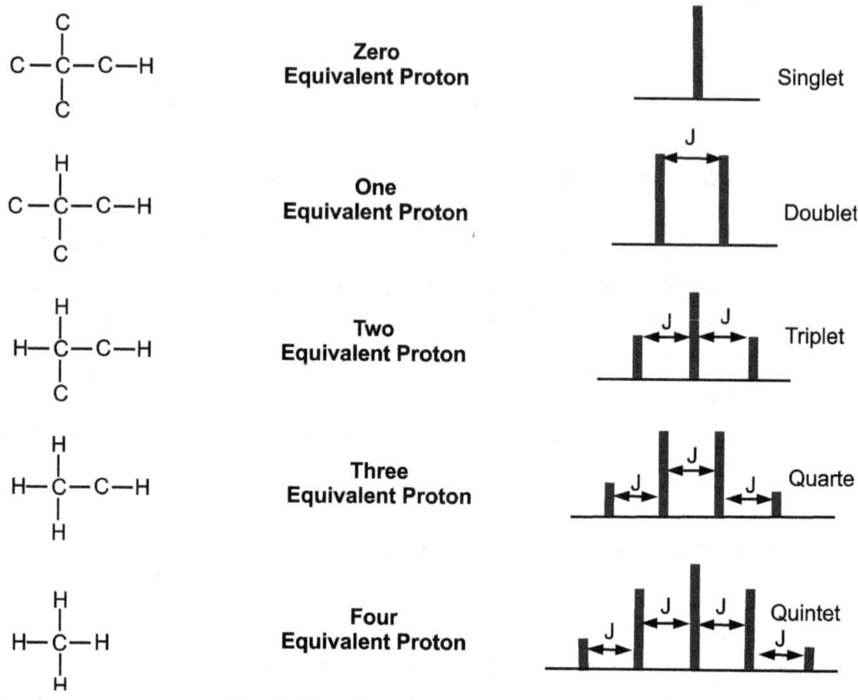

Fig. 5.22 H-NMR splitting patterns

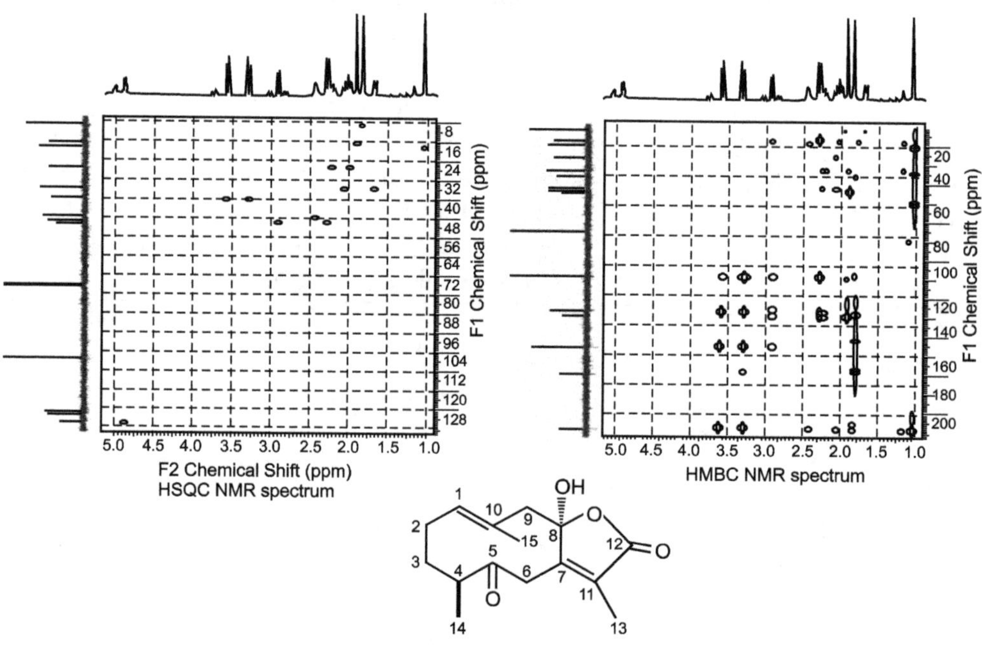

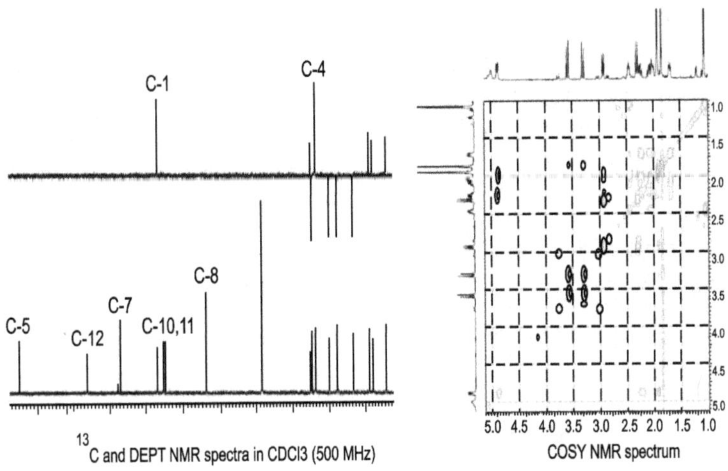

Fig. 5.23 Examples of 2D NMR spectra

Source: PhD thesis of Myint Myint Khine, University of Yangon, Myanmar

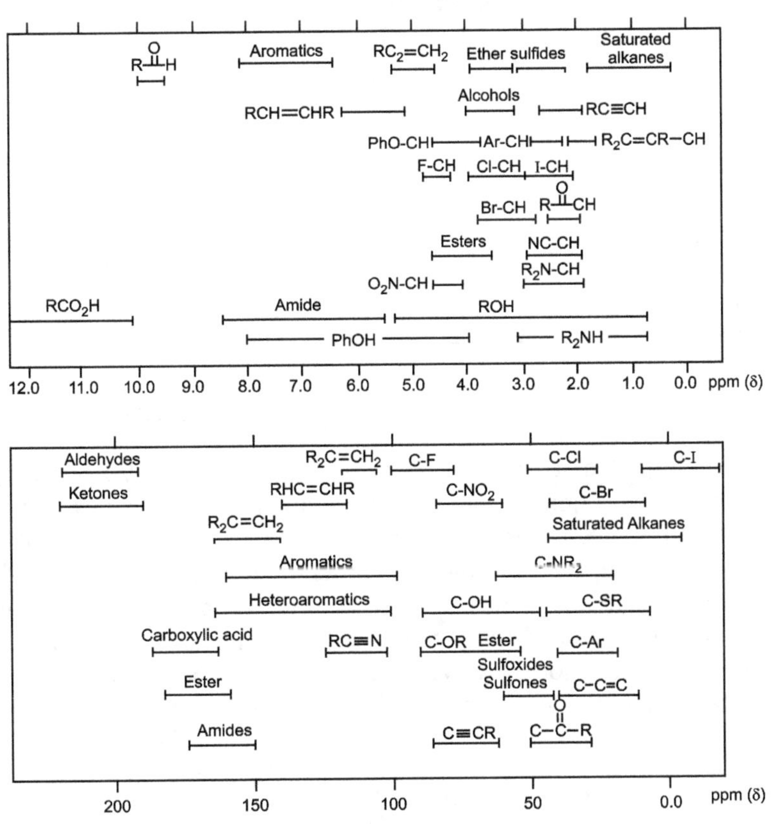

Fig. 5.24 NMR chemical shifts pertaining to various functional groups

Mass Spectroscopy

Mass spectrometry is adifferent spectroscopical technique that identifies and determines the molecular weight as well as structural pattern based on the mass-to-charge ratio of charged particles. The typical instrumentation of mass spectrophotometer has been shown in Figure 5.25 which contains:

1. **Ion source (the heart of this instrument):**A small sample of compound is ionized, usually to cations by loss of an electron.

2. **Mass analyzer:**Ions are sorted and separated according to their mass and charge.

3. **Detector:**Separated ions are then detected and tallied, and the results are displayed on a chart.

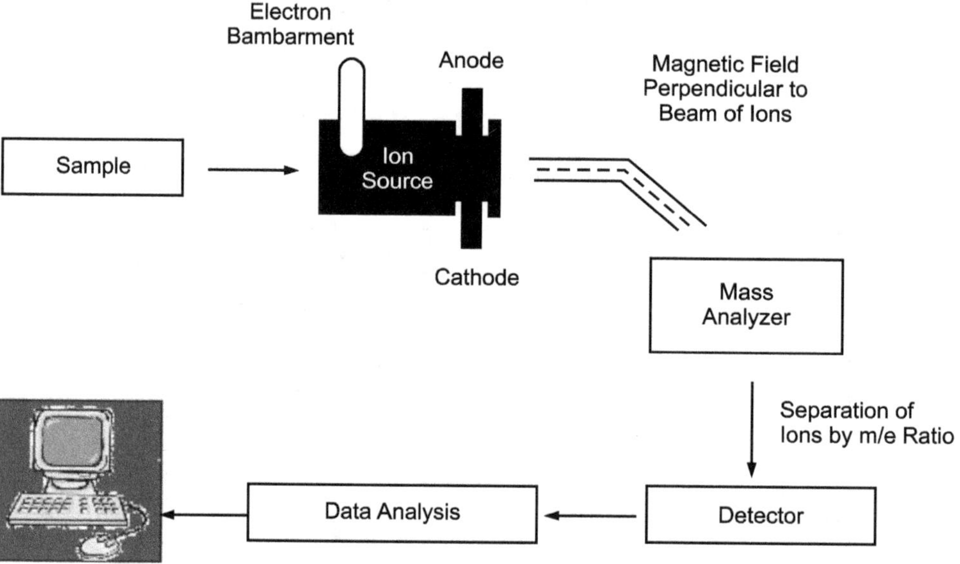

Fig.5.25 Block diagram of mass spectrophotometer

Working: Sample is subjected to ionisation,i.e.,chemicalfragmentation through high energy beam of electrons which forms anions, cations, and a few neutral ions.In mass analyser,cations are accelerated toward other electrodes and are focusedin a beam. This beam of cations is, then, bent due to an external magnetic field which is perpendicular to its direction of motion. Thus, the ions are deflected in an arc whose radius is directly proportionalto the mass of the ion.Changes in the strength of the magnetic field allows ions of different mass to be focused progressively on a detector.As ions are very reactive and short-lived, their formation and manipulation must be conducted in a vacuum.*The formula to calculate the mass to charge rationis*

$$\frac{m}{z} = \frac{B^2 r^2}{2V}$$

Where, V is velocity, B: is magnetic field, r is radius of circular path. B, V and Z are kept constant in the mass spectrophotometer, then mass m is an exclusive function of radius of circular path "r" or vice versa.

Techniques in Mass Spectroscopy

Mass spectrometry is highly sensitive, fast, and selective. Combination of mass spectrometry with HPLC, GC, or with an additional stage of mass spectrometry, known as Tandem mass spectrometry i.e MS/MS, increases the selectivity considerably. Tandem mass spectrometers are instruments that have more than one analyzer, and thus, it can be used effectively for structural and sequencing studies. For high molecular (MW 1000 – 5000) and complex natural products (e.g.,saponins, glycosides, polymers, pigments) FAB, MALDI and ESI techniques should be preferred.

- **APCI:** Atmospheric Pressure Chemical Ionization
- **CI:** Chemical Ionization
- **EI:** Electron Impact
- **FAB:** Fast Atom Bombardment
- **MALDI:** Matrix-Assisted Laser Desorption/Ionization
- **LSIMS:** Liquid Secondary Ion Mass Spectrometry
- **CID:**Collision-Induced Dissociation
- **ESI:**Electro Spray Ionization

Important Terminology in Mass Spectrum

Molecular peak: This is the last peak in the mass spectra which is an indication of molecular weight of molecule. It is obtained by the loss of an electron from the molecule and always denoted with symbolsM+, M+1, M+2.

Parent/Base peak: This is the most intense peak of 100% intensity. Heights of all other peaks are measured with the intensity of this peak.

Fragment ion: Stable and lighter carbocations formed by the decomposition of the molecular ion are called as Fragment ion.

Rearrangement ions: These ions are formed due to the redistribution of fragmented ions, and hence, these are not actual parts of original molecules.

Metastable ions: These ions are formed during decomposition of molecule between the ion source and mass analyzer. Metastable ions play an important role in studying fragmentation pattern.

Multiple charged ions: The doubly, triply charged ions are the most common part of heteroatomic molecules, and these are found at 1/2 or 1/3rd values of mass to charge (m/e) ratio of molecule.

Negative ions: These are less common ions formed during ionization process and of less value in structural elucidation.

Nitrogen rule: This rule is useful to determine the nitrogen content of an unknown compound.If a compound contains an even or odd number of nitrogen atoms (or no nitrogen atoms), its molecular weight will be even or odd respectively.

Ring rule: This rule is useful to determine the number of unsaturated sites of an unknown compound. Number of unsaturated sites is equal to the number of rings present in the molecule plus the number of double bonds plus twice the number of triple bonds.

Table 5.6 Common fragment ions in mass spectrometry and their associated functional groups

Mass	Associated fragments	Associated functional groups
15	CH_3+	Methyl, Alkane
16	O	$-NO_2$, S=O, =N-O
16	NH_2	$ArSO_2NH_2$, $-CONH_2$
17	NH_3	Amine
17	OH	Alcohol
18	H_2O	Ketone Aldehyde, Alcohol
28	CO	Quinone
28	C_2H_4	Aromatic ethyl ether, ethyl ester, propyl ketone
29	CHO	Aldehyde
29	C_2H_5	Ethyl ketone
30	C_2H_6	Alkane
30	OCH_2	Aromatic methyl ether
30	NO, CH_2, $=NH_2$	Aromatic NO_2, amine
31	OCH_3	Methyl ester
31	CH_3OH	Methyl ester
33	HS	Thiol
34	H_2S	Thiol
41	C_3H_5	Propyl ester
41	CH_2CO	Methyl ketone, aromatic ethanaoate
42	C_3H_6	Butyl ketone, aromatic propyl ether
44	CO_2	Ester, anhydride, carboxylic acid
45	COOH	Carboxylic acid
46	C_2H_5OH	Ethyl ester
46	NO_2	Aromatic-NO_2
48	SO	Aromatic sulphoxide
50	C_4H_2	Aryl
77	C_6H_5	Phenyl
83	C_6H_{11}	Cyclohexyl
91	C_7H_7	Benzyl
105	C_6H_5-R, C_6H_4-RR'	Substituted benzene

Table 5.7 Types of Mass Spectrometers and their *m/z* Range

Instrument	*m/z* Range
Magnetic sector	12,000
Quadrupole	4,000
Triple quadrupole	4,000
Time-of-flight (TOF)	>200,000
Fourier transform ion cyclotron resonance (FTICR)	<10,000
Quadrupole time-of-flight (QTOF)	4,000
Time-of-flight - time-of-flight (TOF-TOF)	>10,000

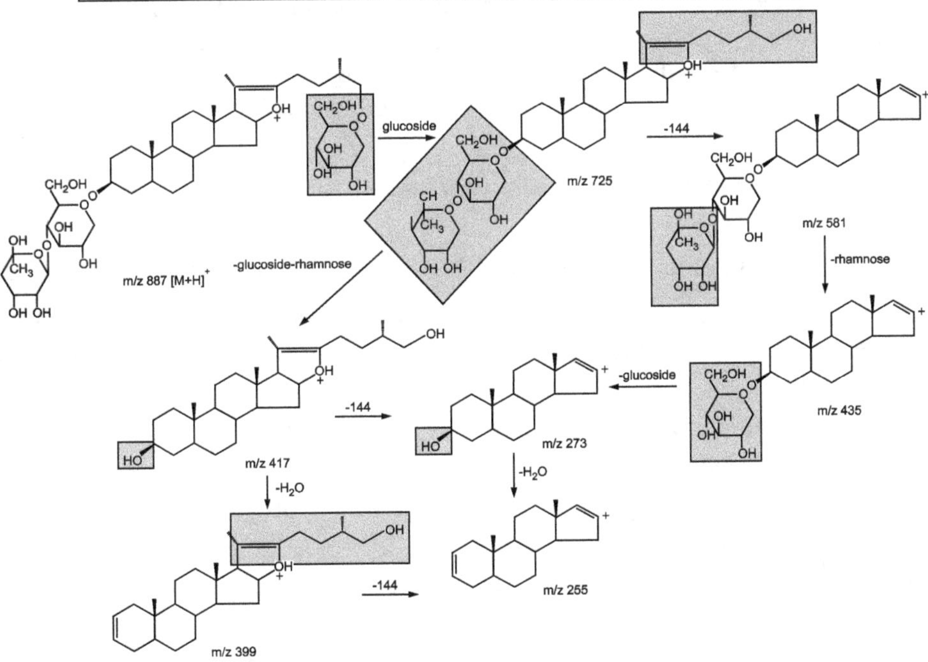

Fig. 5.26 Example of Mass spectrometric fragmentation pattern for stroidal glycosides

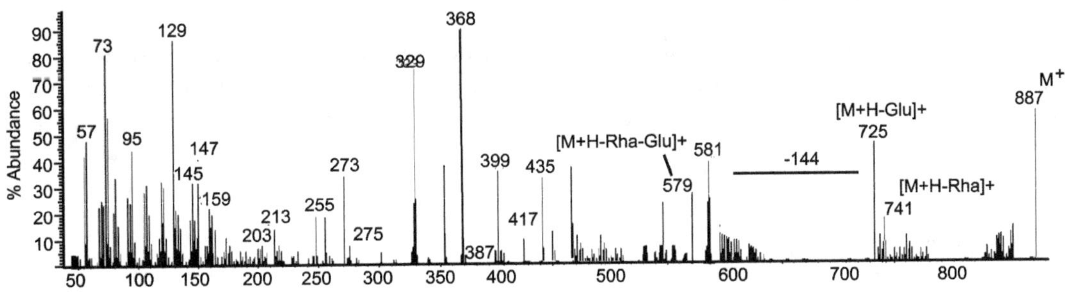

Fig. 5.27 ESI-MS/MS fragmentation pattern for one steroidal

saponin **Source:** F. Liang et. al. 2002

Applications of Spectroscopy

Spectroscopic Elucidation of Phytoconstituents

The molecular spectra contain important information about the structure of a molecule. The information available from different types of molecular spectra can be used for the structure elucidation of the organic molecules. This is done by co-rrelating the spectral signals to different structural units present in the molecule. Once the correlation between different spectral features and the structure of the molecule is understood,it is possible to predict the important spectral details for a given molecule from its structure. In this context, we will use the information available from UV-Vis, IR, ^{1}HNMR, ^{13}CNMR, and mass spectrometry.

UV spectrum is used to predict whether compound is aromatic or aliphatic and saturated or unsaturated. Fingerprint region of IR spectrum gives details of functional groups present in a molecule. NMR yields information about quality and quantity of hydrogen and carbon atoms in the structure. Mass spectroscopy is useful to determine molecular weight and fragmentation pattern of a molecule.

In this way, every spectroscopic data is very useful for detail construction of molecule's architecture.

Thus following important points cane be considered while elucidating molecule's structure:

- Calculate Index of Hydrogen Deficiency (IHD) from molecular formula because once, the molecular formula of an unknown compound is known, structural information can be elucidated by IHD.

- List important structural information available from UV-VIS, IR, ^{1}HNMR, ^{13}C NMR, and mass spectra of a molecule.

- Correlate the signals and their intensities in different types of spectra of a molecule to various structural units present in it.

- Integrate the information available from different types of spectra to elucidate the structure of a molecule.

- Predict the spectral data for a compound from its structural formula.

Index of Hydrogen Deficiency (IHD)

IHD is a count of how many molecules of hydrogen need to be added to a structure in order to obtain the corresponding saturated, acyclic species.It indicates how many rings and multiple bonds are present in the structure.If the molecular formula is $C_cH_hN_nO_oX_x$, then the equation to calculate IHD is

$$IHD = 0.5 \times [2c+2-h-x+n]$$

Where, c=carbon,

      h=hydrogen,

      x=halogen and

      n=nitrogen

- If molecule contains oxygen, ignore number of oxygen while determining IHD.
- If molecule contains halogen, then consider number of halogen as number of hydrogen.
- IHD gives idea about what types of structural units are possible and which structure exactly fits to the molecular formula.
- The unknown compound must have at least one double bond or one ring.

SummerisedApplications of spectroscopy in Pharmacognosy		
Type of spectroscopy	**Specific Applications**	**General applications**
Ultraviolet (UV)-visible spectroscopy	To determine un-saturation, quantitative estimation To determine UVmax	To know purity of phytochemicals, to identify impurities, to differentiate isomers
Infra-red (IR) spectroscopy	To identify functional groups present in phytochemicals	
Nuclear magnetic resonance (NMR) spectroscopy	To determine quality and quantity of nuclei like H, C, N, P, F	
Mass spectroscopy	To determine molecular weight and fragmentation pattern of phytochemicals	

Subjective Questions

1. How to extract crude drugs? What are different methods of extraction of crude drugs?
2. What are advantages and disadvantages of microwave extraction or Supercritical fluid extraction method?
3. How different chromatographic techniques are useful in Pharmacognosy?
4. How different spectroscopic techniques are useful in Pharmacognosy?
5. Absorption, emission, fluorescence, phosphorescence, luminescence, photoluminescence, chemiluminescence, thermal relaxation, intersystem crossing, and internal conversion are commonly used terms in spectroscopy. Define each term. How are the terms related?
6. How different electrophoresis techniques are useful in Pharmacognosy?
7. What is principle of Chromatography?
8. What is principle of UV-Spectroscopy?
9. What is principle of Electrophoresis?
10. What is difference between TLC and HPTLC?
11. What is difference between Column chromatography and HPLC?.
12. What are general techniques involved in Planar chromatography?
13. What are general techniques involved in column chromatography?

14. What are the functions of different chemicals used in Electrophoresis?

15. What is Rf and Rt?

Multiple Choice Questions (MCQs)

1. Continuous hot percolation is principle of

    Super critical fluid extraction

    Microwave assisted extraction

    c. Soxhlet extraction

    d. Maceration

2. Which principle is applicable for selection of solvents for extraction process?

    a. Like cures like

    b. Like repels like

    c. Like dissolves like

    d. None of the above

3. ....is the most commonly used frequency in commercial microwave instruments

    a. 2.5 GHz

    b. 3 GHz

    c. 2.45 GHZ

    d. 5 GHz

4. In super critical fluid extraction, Critical temperatures and pressure of carbon dioxide should be

    a. 20.9°C and 72.8 atm respectively

    b. 40.9°C and 74.8 atm respectively

    c. 30.9°C and 73.8 atm respectively

    d. 50.9°C and 75.8 atm respectively

5. Which one of the modern extraction technique is highly suitable for heat labile compounds?

    a. Soxhlet extraction

    b. Microwave assisted extraction

    c. Super critical fluid extraction

    d. Ultrasound assisted extraction

6. At industrial scale, caffeine is extracted by using

    a. Soxhlet extraction

    b. Microwave assisted extraction

    c. Super critical fluid extraction

    d. Ultrasound assisted extraction

7. Which one of the following spectroscopic technique is used to understand conjugated pi-electron system in a phytochemical?

    a. IR Spectroscopy

    b. UV visible spectroscopy

    c. MS spectroscopy

    d. NMR

8. Determination of molecular structure, molecular weight and presence of isotop patterns for Cl and Br can be performed by

    a. Mass spectroscopy

    b. NMR spectroscopy

    c. IR spectroscopy

    d. Both a and b

9. Which one of the following technique is termed as 'Time domain spectrometry'?

    a. IR spectroscopy

    b. FTIR spectroscopy

    c. NMR spectroscopy

    d. Mass spectroscopy

10. Number of carbon and hydrogen atoms present in a compound can be determined by using
    a. IR spectroscopy
    b. FTIR spectroscopy
    c. NMR spectroscopy
    d. Mass spectroscopy

11. In IR spectroscopy, the region between wavelength $1500 - 500$ cm^{-1} is called finger print region because
    a. Fingerprints can be detected in this region
    b. Each different compound produces a different pattern of trough in this part of spectrum
    c. Both a and b
    d. None of these

12. Which one of the following is NOT a non chromatographic technique of purification?
    a. Fractional crystallization
    b. Sublimation
    c. Distillation
    d. electrophoresis

13. The technique in which antimicrobial activity of a bioactive compound is detected by using TLC chromatography is known as
    a. Zone of inhibition
    b. Autobiography
    c. Bio autography
    d. None

14. The HPLC system is called as isocratic elution system when
    a. Volume of the mobile phase changes during separation
    b. Composition of mobile phase remains constant during separation
    c. Composition of mobile phase changed during separation
    d. Volume of the mobile phase remains constant

15. In HPLC analysis, Tailing factor (T) describes
    a. Peak symmetry in chromatogram
    b. Peak asymmetry in chromatogram
    c. Peak height in chromatogram
    d. Number of peaks in chromatogram

16. The delay time (t0) in HPLC analysis indicates
    a. The time required by non retarded compound to travel from injection slit to detector
    b. The time required by compound from the moment of injection until the time of detection
    c. Both a and b
    d. None

17. During the analysis of compound by using chromatography, which factor is substance specific and should always provide the same value under the same condition?

    a. Refractive index
    b. Retention time
    c. Retention factor
    d. Both b and c

18. Which one of the following analytical methods is used to separate macromolecules based on size?

    a. HPLC
    b. TLC
    c. HPTLC
    d. Electrophoresis

19. Which one of the following electrophoresis technique is applicable for analysis of both charged and non charged molecules?

    a. SDS – PAGE
    b. Capillary electrophoresis
    c. Gel electrophoresis
    d. None

20. Which one of the following electrophoresis technique is suitable for analysis of phytochemicals?

    a. Capillary electrophoresis
    b. SDS –PAGE electrophoresis
    c. Gel electrophoresis
    d. Both a and c

21. Which technique separates charged particles using electric field?

    a. Hydrolysis
    b. Electrophoresis
    c. Protein
    d. denaturing synthesis

22. Electrophoresis was developed by:

    a. Tswett
    b. Tsvedberg
    c. Tiselius
    d. Sanger

23. The speed of migration of ions in electric field depends upon:

    a. Shape and size of molecule
    b. Magnitude of charge and shape of molecule
    c. Magnitude of charge shape and mass of molecule
    d. Magnitude of charge and mass of molecule

24. Which of the following statements is true about migration of biomolecules?

    a. The rate of migration is directly proportional to the resistance of medium
    b. Rate of migration is directly proportional to current
    c. Low voltage is used for separation of high mass molecules
    d. Rate of migration is inversely proportional to current

25. What does the electrophoresis apparatus consist of?

    a. Gel, buffer chamber and fire pack
    b. Buffer chamber and electrophoresis unit
    c. Electrophoresis unit and gel separator
    d. Power pack and electrophoresis unit

26. If proteins are separated according to their electrophoretic mobility then the type of electrophoresis is:

    a. SDS PAGE
    b. Affinity Electrophoresis
    c. Electro focusing
    d. Free flow electrophoresis

27. The electrophoretic mobility denoted as μ is mathematically expressed as:

    a. VE
    b. E/V
    c. 1/EV
    d. V/E

28. Which of the following factors does not influence electrophoretic mobility?

    a. Molecular weight
    b. Shape of molecule
    c. Size of molecule
    d. Stereochemistry of molecule

29. When is electrophoresis not used?

    a. Separation of proteins
    b. Separation of amino acids
    c. Separation of Lipids
    d. Separation of nucleic acids

30. What cannot be a reason for using electrophoresis?

    a. Comparing two sets of DNA
    b. Organizing DNA by shape of backbone
    c. Organizing DNA fragments from largest to smallest
    d. Organizing DNA in order we can see

Answer Key

1.c	2. c	3. a	4. c	5. c	6. c	7. b	8.a	9.b	10. c
11. c	12. d	13. c	14. b	15. a	16. a	17. d	18. d	19. b	20. a
21.b	22.c	23.b	24.b	25.d	26. a	27. d	28. d	29. c	30. b

Further Reading

1. A.N.M. Alamgir. Therapeutic Use of Medicinal Plants and Their Extracts: Volume 1

2. Alexander I. Gray, Satyajit D. Sarker, Zahid Latif. Natural Products Isolation. Humana Press. 2006

3. Alice Kurian, M. Asha Sankar Medicinal Plants. New India Publishing Agency. 2007

4. Ashutosh Kar. Pharmacognosy and Pharmacobiotechnology. New Age International (P) Limited. 2003

5. Ayurvedic pharmacopoeia of India Part-I vol.I, 2001.

6. Bernard Fried. Handbook of Thin-Layer Chromatography. Taylor & Francis. 2003

7. Biren Shah, Avinash Seth. Textbook of Pharmacognosy and Phytochemistry. Elsevier Health Sciences. 2014

8. C. S. Shah, J. S. Qadry. A Textbook of Pharmacognosy. Messrs B.S. Shah 1971

9. C.K. Kokate, Purohit, Gokhlae. Text book of Pharmacognosy, 37th Edition, Nirali Prakashan, Pune. 2007

10. Chi-Tang Ho, Fereidoon Shahidi. Phytochemicals and Phytopharmaceuticals. AOCS Press. 2000

11. Corrado Tringali. Bioactive Compounds from Natural Sources-Isolation, Characterization and Biological Properties. Taylor & Francis. 2003.

12. Debra K.W. Topham, Mark S. Meskin, Stanley T. Omaye, Wayne R. Bidlack. Phytochemicals as Bioactive Agents. Taylor & Francis.2000

13. Deore SL, Khadabadi SS, Baviskar BA. Pharmacognosy and Phytochemistry-A Comprehensive Approach. PharmMed Press, Hyderabad. 2nd Edition, 2018.

14. Deore SL. Pharmacognosy and Phytochemistry: A Companion Handbook. PharmMed Press, Hyderabad. . 2nd Edition, 2017.

15. Derek J. Chadwick, Joan Marsh. Bioactive Compounds from Plants. Wiley. 2008

16. Elke Hahn-Deinstrop. Applied Thin-Layer Chromatography-Best Practice and Avoidance of Mistakes. Wiley. 2007

17. Finian J. Leeper, John C. Vederas. Biosynthesis Springer, London. 2000

18. Fraiture MA, Herman P, Taverniers I, De Loose M, Deforce D, Roosens NH. Current and new approaches in GMO detection: challenges and solutions. Biomed Res Int. 2015;2015:392872.

19. GS Kumar. KN Jayaveera. A Textbook of Pharmacognosy and Phytochemistry. S Chand & Company Limited. India. 2014.

20. Gunnar Samuelsson. Drugs of Natural Origin-A Textbook of Pharmacognosy. Apotekarsocieteten. 1999

21. H. Ansari. Essentials of Pharmacognosy. Second edition, Birla publications, New Delhi, 2007

22. Hany El-Shemy. Aromatic and Medicinal Plants-Back to Nature. IntechOpen.2017

23. Indian Pharmacopeia 2018, Ghaziabad: Indian Pharmacopeia Commission; 2018.

24. James Bobbers, Marilyn KS, VE Tylor. Pharmacognosy &Pharmacobiotechnology. Williams & Wilkins. 1996.

25. Jean Bruneton. Pharmacognosy, Phytochemistry, Medicinal Plants. Technique & Documentation. 1999

26. Jeffrey B. Harborne. Phytochemical Methods-A Guide to Modern Techniques of Plant Analysis. Springer Netherlands. 2012

27. John T. Arnason, John T. Romeo, Rachel Mata. Phytochemistry of Medicinal Plants. Springer US. 2013

28. Joseph Sherma, Monika Waksmundzka-Hajnos, Teresa Kowalska. Thin Layer Chromatography in Phytochemistry. CRC Press. 2008

29. Joseph Sherma, Monika Waksmundzka-Hajnos. High Performance Liquid Chromatography in Phytochemical Analysis. CRC Press. 2010

30. Juliana M Prado, Mauricio A Rostagno. Natural Product Extraction-Principles and Applications. Royal Society of Chemistry. 2013

31. K. Mangathayaru. Pharmacognosy: An Indian perspective. Pearson Education India. 2013

32. Kaliya.A. Text Book of Industrial Pharmacognosy. CBS Publishers & Distributors, Delhi. 2009

33. Kendall Jefferson. Pharmacognosy and Phytotherapy. Foster Academics.2019

34. Khadabadi SS, Deore SL, Baviskar BA. Experimental Phytopharmacognosy. Nirali prakashan, Pune. 1st Edition, 2019.

35. Luqi Huang. Molecular Pharmacognosy. Springer Netherlands. 2012

36. M.R.F. Ashworth. Egon Stahl. Thin-Layer Chromatography-A Laboratory Handbook. Springer Berlin Heidelberg. 2013

37. Mallappa Kumara Swamy. Plant-derived Bioactives-Production, Properties and Therapeutic Applications. Springer Singapore. 2020

38. Michael Heinrich, Elizabeth M. Williamson, Joanne Barnes, Simon Gibbons, Jose Prieto-Garcia Fundamentals of Pharmacognosy and Phytotherapy E-Book. Elsevier Health Sciences. 2017

39. Michael Heinrich, Joanne Barnes, Simon Gibbons. Fundamentals of Pharmacognosy and Phytotherapy. Churchill Livingstone/Elsevier. 2012

40. Mohammad Ali. Pharmacognosy and Phytochemistry, CBS Publishers & Distribution, New Delhi.

41. N P S Sengar, Ashwini Singh, Ritesh Agrawal. A Textbook of Pharmacognosy. PharmaMed Press. 2018

42. N. Raaman. Phytochemical Techniques. New India Publishing Agency. 2006

43. Paul M Dewick. Medicinal Natural Products-A Biosynthetic Approach Second EditionJohn Wiley & Sons Ltd, London. 2002

44. Pharmacognosy- Volume 1. Springer International Publishing. 2017

45. Rangari VD. Pharmacognosy& Phytochemistry. Career Publication, Nashik. 2008

46. Raphael Ikan. Natural Products-A Laboratory Guide. Elsevier Science. 2013

47. Reiner Westermeier. Electrophoresis in Practice-A Guide to Methods and Applications of DNA and Protein Separations. Wiley. 2016

48. Richard J. P. Cannell. Natural Products Isolation. Humana Press. 1998

49. S. S. Agarwal, M. Paridhavi. Herbal Drug Technology. Universities Press. 2012

50. S. S. Handa. Pharmacognosy. Vallabh Prakashan, New Delhi. 1989

51. Saikat Sen, Raja Chakraborty Herbal Medicine in India-Indigenous Knowledge, Practice, Innovation and Its Value. Springer Singapore. 2019

52. Satyajit D. Sarker, Lutfun Nahar. Computational Phytochemistry. Elsevier. 2018

53. Simone Badal Mccreath. Rupika Delgoda. Pharmacognosy-Fundamentals, Applications and Strategies. Elsevier Science. 2017

54. Stefan Berger, Dieter Sicker. Classics in Spectroscopy-Isolation and Structure Elucidation of Natural Products. Wiley. 2009

55. T. C. Denston. A Textbook of Pharmacognosy. Read Books. 2012

56. T.RabilloudProteome Research: Two-Dimensional Gel Electrophoresis and Identification Methods. Springer.2000

57. T.A. Scott, Sabine Bladt, Eva M. Zgainski. Plant Drug Analysis-A Thin Layer Chromatography Atlas. Springer Berlin Heidelberg. 2013

58. The British Pharmacopeia. London: Medicines and Healthcare Products Regulatory Agency; 1993.

59. United States Pharmacopoeia and National Formulary, USP 25 NF 19/National Formulary 20, Rockville, MD, U. S. Pharmacopoeial Convention, Inc. 2002.

60. Venketeshwer Rao, Leticia G. Rao. Phytochemicals Isolation, Characterisation and Role in Human Health. IntechOpen. 2015

61. W.C.Evans, Trease and Evans Pharmacognosy, 16th edition, W.B. Sounders & Co., London, 2009.

Part – II

Practical Manual

Know Subject: Pharmacognosy

The term 'pharmacognosy' (combination of two Greek words i.e. *pharmakon* means drug and *gnosis* means knowledge) means acquiring knowledge of drugs was coined in 1815 by C. A. Seydler, German medical student in his thesis title *"Analyetica Pharmacognostica"*. Pharmacognosy is defined as scientific and systematic study of structural, physical, chemical and biological characters of crude drugs along with history, method of cultivation, collection and preparation for the market. The American Society of Pharmacognosy defines pharmacognosy as "the study of the physical, chemical, biochemical and biological properties of drugs, drug substances or potential drugs or drug substances of natural origin as well as the search for new drugs from natural sources. It is also called as study of crude drugs.

Thus pharmacognostical studies of plant drugs involves study of synonyms, vernacular names, Biological sources, distribution, morphology, histology, chemistry, qualitative test, various physicochemical tests, pharmacological actions along with commercial varieties, substitutes, adulterants and any other quality control parameters of the drugs.

However, this subject is as old as pharmacy and mankind evolution; recently it is evolved as a multidisciplinary subject focusing many modern disciplines like ethanobotany, ethanopharmcology, phytotherapy, phytochemistry, chemo-taxanomy, biotechnology, clinical trials, herbal drug interaction and even novel drug delivery systems like phytosomes rather only botanical and taxanomical descriptions. Recent advances in extraction methods, analytical hyphe-nated techniques, screening methods continues to hasten major changes in this subject. Modernization of conventional and/or traditional dosage forms is opening doors to industrial Pharmacognosy.

Due to most recent technologies and innovative chemical concepts, many new drugs or drug candidates still originated from natural products or derivatives thereof. Even in this era of nanotechnology, natural drugs are important part of primary health care which is giving pharmacognosy professionals new possibilities to exploit the huge diversity designed and generated by nature.

There is a shortage of established scientists engaged in pharmacognosy research, which tends to involve subject matter beyond the conventional scientist's knowledge base. Hence, actual secret of opportunities in pharmacognosy research is that only the tip of the iceberg seems to have been discovered yet.

Following materials are required for Pharmacognosy laboratory work.

- Napkin
- Needle
- Filter paper

- Camel hair brushes
- Stains
- Watch glass
- A sharp razor blades
- Forceps
- Micro-slide
- Cover slip

Students shall read the points given below for understanding theoretical concepts and practical applications.

1. Students should wear white Apron, Cap, Mask, Gloves and Slipper before entering in to laboratory.

2. Students should keep their belongings in locker which are not required during practical like bag, Extra files etc.

3. Students should always carry Laboratory Manual, rough notebook, and practical requirements without fail.

4. Listen carefully to the lecture given by teacher about importance of subject, curriculum philosophy, graphical structure, skills to be developed, information about equipment, instruments, procedure, method of continuous assessment, tentative plan of working laboratory and total amount of work to be done in a year.

5. Students should perform the practical only at the place which allocated to him/her. (No change can be done without permission of subject teacher)

6. Students shall undergo study visit of laboratory for types of equipment, instruments, material to be used, before performing experiment

7. Read write up of each experiment to be performed, a day in advance.

8. Organize the work in the group and make a record of all observations.

9. Understand the purpose of experiment and its practical applications.

10. Write the answer of the questions allotted by teacher during practical hours if possible or afterwards, but immediately.

11. Students should not hesitate to ask any difficulty faced during conduct of practical.

12. The students shall study all the questions given in the laboratory manual and practice to write the answers to these questions

13. Students shall develop maintenance skill as expected by the industries.

14. Students should develop the habits of pocket discussion, group discussion related to the experiments so that exchanges of knowledge, skills could take place.

15. Students shall attempt to develop related hands on skills and gain confidence.

16. Students shall visit nearby workshops, workstation, industries, technical exhibitions, trade fair etc. even not included in the lab manual. In short, students should have exposure to the area of work right in the student's hood.

17. Students shall insist for the completion of recommended laboratory work, industrial visits, answers to the given questions, etc

18. Students shall develop habits of evolving more ideas, innovations skills etc. than included in the scope of manual

19. Students shall develop technical magazines, proceedings of seminars, refers websites related to the scope of the subjects and update their knowledge and skills.

20. Students should develop the habit of not to depend totally on the teachers but to develop self learning techniques

21. Students should develop the habit to react with the teacher without hesitation with respect to the academic involved.

22. Students should develop the habit to submit the practical exercise continuously and progressively on the scheduled dated and should get the assessment done

23. Student should be well prepared while submitting the write up of the experiments. This will develop the continuity of the studies and he will be over laded at the end of the term.

24. Students should clean platform before leaving the laboratory.

Index

Sr. No	Aim	Date	Page	Grade/ Marks	Signature
1	To study morphological, histological, powder microscopical and chemical characteristics of crude drug- *Cinchona*				
2	To study morphological, histological, powder microscopical and chemical characteristics of crude drug- *Clove* https://youtu.be/x0anOwHH0Nc				
3	To study morphological, histological, powder microscopical and chemical characteristics of crude drug- *Coriander* https://youtu.be/jK-91Y-40uM				
4	To study morphological, histological, powder microscopical and chemical characteristics of crude drug- *Ephedra* https://youtu.be/cLYT-9eK2dk				
5	To study morphological, histological, powder microscopical and chemical characteristics of crude drug- *Fennel* https://youtu.be/wVtOQfekdQ0				
6	To study morphological, histological, powder microscopical and chemical characteristics of crude drug- *Senna*				
7	To study morphological, histological, powder microscopical and chemical characteristics of crude drug- *Cassia/Cinnamon* https://youtu.be/cLYT-9eK2dk				
8	To isolate caffeine from tea dust and confirm by chemical tests. https://youtu.be/DflsWOp9TC4				
9	To extract eucalyptus oil by Clevenger apparatus (Hydro distillation) https://youtu.be/nZFjmbRrkek				
10	To perform TLC of extracted eucalyptus oil. https://youtu.be/eP7fNUNrbpU				

11	To isolate Sennosides from Senna and confirm by chemical tests. https://youtu.be/RMy8AHO10pg				
12	To isolate Disogenin from Dioscorea/Methi seeds and confirm by chemical tests. https://youtu.be/65USUQ0D8kI				
13	To isolate Atropine from Belladonna/Datura and confirm by chemical tests.				
14	To analyse crude drugs (Aloe, Benzoin, Myrrh, Asafoetida, Colophony) by chemical tests				
15	To perform separation of sugars by Paper chromatography https://youtu.be/miiXl8lDKwI				
	Further Reading				

Aim 01: **To study morphological, histological, powder microscopical and chemical characteristics of crude drug-** *Cinchona*

Requirements: Microscopes, slides, test tubes, all pharmacognosy reagents and chemical test reagents.

Theory

Cinchona: Cinchona (Peruvian bark, Jesuit's bark) is dried root or stem bark of *Cinchona calisaya, C.officinalis, C.ledgeriana, C.succirubra*, Family-Rubiaceae. It contains quinoline alkaloids like quinine, quinidine, cinchonine, cinchonidine, cupreine, hydroquinine. It is used as antimalarial and bitter tonic.

Experimental Morphological Study:

Color	
Odor	
Taste	
Size	
Shape	
Extra feature	

Microscopical study:

Transverse section of bark shows cork composed of thick-walled polygonal cells with brown contents, cortex of parenchymatous cells with scattered microprism crystals of calcium oxalate and starch grains, Sclereids and stone cells are absent in most of the *Cinchona* species, abundant thick, lignified, yellowish, fragmented and forked fibers, abundant parenchyma of the phloem and medullary rays and sieve tubes with companion cells are the characteristic features.

Labeled diagram of Morphological characters of crude drug

Labeled photograph of TS of crude drug

Powder Microscopical characters with diagram/ photographs and description

Micro-chemical study:

Micro-chemical tests	Name of test	Observation	Inference
Lignin			
Starch grain			
Calcium crystals			
Oil (Volatile/non-volatile)			
Major secondary metabolite			

Powder Characteristics:

Color	
Odor	
Taste	
Powder Microscopical features	

Chemical study:

Chemical class	Name of test	Observation	Inference
Alkaloid test			
Tannins test			

Result: From morphological, microscopical, micro chemical, powder and chemical study it is confirmed that given crude drug is ....

Questions

1. What are the special morphological characters of bark cinchona.
2. Which is confirmatory chemical test for cinchona alkaloids
3. How much minimum percentage of total alkaloids present in cinchona bark?
4. Which type of calcium crystals and xylem vessels are present in cinchona bark?
5. What is lenticle and lichen?
6. Draw neat labeled diagram of TS of cinchona bark.
7. Draw neat labeled diagram of cinchona powder characters.

Aim 02: To study morphological, histological, powder microscopical and chemical characteristics of crude drug- *Clove*

Requirements: Microscopes, slides, test tubes, all pharmacognosy reagents and chemical test reagents.

Theory

Clove: Clove (Clove-flower, Clove-buds, Caryophyllum) is dried flower buds of *Eugenia caryophyllus* Family Myrtaceae. It contains Eugenin ,Eugenol acetate, Caryophyllin, Ester, ketone and alcohol, eugenol, ester eugenin. It is used as a dental analgesic, carminative, flavourant, and stimulant, aromatic and antiseptic.

Experimental Morphological Study

Color	
Odor	
Taste	
Size	
Shape	
Extra features	

Microscopical study:

Transverse section shows epidermis composed of small, polygonal cells with numerous circular anomocytic stomata and brown ovoid oil glands, cluster crystals of calcium oxalate. Hypanthium is lined with thick cuticle, is collenchymatous with yellowish-brown parenchyma of the hypanthium with schizolysigenous oil glands and small cluster crystals of calcium oxalate.Presence of fragments of the aerenchyma of the hypanthium, fragments of the fibrous layer of small cells of the anther, abundant small, biconvex pollen grains, sclereids from the stalk with brown contents in lumen and central columella of thick parenchymatous cells with calcium oxalate crystals are the main distinguishing features.

Labeled diagram of Morphological characters of crude drug

Labeled photograph of TS of crude drug

Powder Microscopical characters with diagram/ photographs and description

Micro-chemical study

Micro-chemical tests	Name of test	Observation	Inference
Lignin			
Starch grain			
Calcium crystals			
Oil (Volatile/non-volatile)			
Major secondary metabolite			

Powder Characteristics:

Color	
Odor	
Taste	
Powder Microscopical features	

Chemical study:

Chemical class	Name of test	Observation	Inference
Essential oil test			
Tannins test			

Result: From morphological, microscopical, micro chemical, powder and chemical study it is confirmed that given crude drug is ……………

Questions

1. What is hypanthium?
2. Draw neat labeled diagram showing morphological characters of clove?
3. Which part of clove contains oil glands?
4. Which part of clove contains calcium crystals?
5. Which is confirmatory chemical test for clove major chemical constituent?
6. How much minimum percentage of total volatile oil present in clove?
7. Draw neat labeled diagram of TS of clove hypanthium and through ovary
8. Draw neat labeled diagram of clove flower bud powder characters.

Aim 03: To study morphological, histological, powder microscopical and chemical characteristics of crude drug- *Coriander*

Requirements: Microscopes, slides, test tubes, all pharmacognosy reagents and chemical test reagents.

Theory

Coriander: Coriander (Coriander fruits) is dried ripe fruits of *Coriandrum sativum*, family Umbeliferae. It contains Pinene, Geraniol, Coriandrol, Coriandryl acetate, L-borneol, cineol. It is used as aromatic, flavourant, carminative, and stimulant.

Experimental Morphological study

Color	
Odor	
Taste	
Size	
Shape	
Extra feature	

Microscopical study:

Transverse section shows epicarp composed of thin-walled polygonal cells with prisms of calcium oxalate; stomata and brown fragments of the vittae; abundant sclerenchyma of the mesocarp, thin-walled, lignified cells forming parquetry layer of endocarp, sclereids occur adjacent to the endocarp, thin-walled polygonal and brown cell layer of testa, and endosperm composed of thick-walled cells with aleurone grains and microrosette crystals of calcium oxalate.

Labeled diagram of Morphological characters of crude drug

Labeled photograph of TS of crude drug

Powder Microscopical characters with diagram/ photographs and description

Micro-chemical study

Micro-chemical tests	Name of test	Observation	Inference
Lignin			
Starch grain			
Calcium crystals			
Oil (Volatile/non-volatile)			
Major secondary metabolite			

Powder Characteristics:

Color	
Odor	
Taste	
Powder Microscopical features	

Chemical study:

Chemical class	Name of test	Observation	Inference
Essential oil test			
Tannins test			

Result: From morphological, microscopical, micro chemical, powder and chemical study it is confirmed that given crude drug is ....

Questions

1. What is vittae?

2. Which is confirmatory chemical test for coriander major chemical constituent?

3. Which is major chemical constituent of Coriander and what are it's commercial uses?

4. Draw neat labeled diagram of TS of coriander.

5. Draw neat labeled diagram of coriander powder characters.

6. Which type of calcium crystals is present in coriander fruit?

Aim 04: To study morphological, histological, powder microscopical and chemical characteristics of crude drug- *Ephedra*

Requirements: Microscopes, slides, test tubes, all pharmacognosy reagents and chemical test reagents.

Theory

Ephedra: Ephedra (Ma huang) is dried stem of *Ephedra gerardiana, E.equeisetina, E. sinica* of family Ephedraceae. It contains amine alkaloids ephedrine, nor ephedrine, n-methyl ephedrine, Pseudoephedrine. It is used as sympathomimetic, antiasthmatic and also useful in treatment of hay fever.

Experimental Morphological study:

Color	
Odor	
Taste	
Size	
Shape	
Extra feature	

Microscopical study:

Transverse section shows epidermal layer of quadrangular parenchymatous cells and shrinked stomata, chlorenchymatous cortex consisting calcium oxalate crystals, spongy parenchyma, slightly lignified mesocarp fibers, pericyclic fibers, open and collateral vascular bundles and pith composed of large cells containing mucilage.

Labeled diagram of Morphological characters of crude drug

Labeled photograph of TS of crude drug

Powder Microscopical characters with diagram/ photographs and description

Micro-chemical study

Micro-chemical tests	Name of test	Observation	Inference
Lignin			
Starch grain			
Calcium crystals			
Oil (Volatile/non-volatile)			
Major secondary metabolite			

Powder Characteristics:

Color	
Odor	
Taste	
Powder Microscopical features	

Chemical study:

Chemical class	Name of test	Observation	Inference
Alkaloid test			
Tannins test			

Result: From morphological, microscopical, micro chemical, powder and chemical study it is confirmed that given crude drug is ....

Questions

1. Which is confirmatory chemical test for ephedra major chemical constituent?
2. Which is major chemical constituent of ephedra and what are it's commercial uses?
3. Draw neat labeled diagram of TS of ephedra stem.
4. Draw neat labeled diagram of ephedra powder characters.
5. Which type of calcium crystals and xylem vessel is present in ephedra stem?

Aim 05: To study morphological, histological, powder microscopical and chemical characteristics of crude drug- *Fennel*

Requirements: Microscopes, slides, test tubes, all pharmacognosy reagents and chemical test reagents.

Theory

Fennel: Fennel (Fennel fruits, Fructus foeniculum) is dried ripe fruits of *Foeniculum vulgare* of family Umbeliferae. It contains volatile oil principles like fenchone, anethol, ketone, phellandrene, limonene, methyl charvicol and anisicaldehyde. It is used as carminative, flavourant, stimulant and expectorant.

Experimental Morphological study:

Color	
Odor	
Taste	
Size	
Shape	
Extra feature	

Microscopy

Transverse section shows epicarp composed of polygonal, thin walled, smoothly cuticulised cells and stomata, mesocarp composed of reticulate, sub rectangular parenchyma cells, endocarp composed of layer of a thin-walled, lignified and elongated cells, endosperm composed of moderately thick-walled cells containing aleurone grains and micro rosette crystals of calcium oxalate, lignified fibro-vascular tissue, parquetry layer and vittae.

Labeled diagram of Morphological characters of crude drug

Labeled photograph of TS of crude drug

Powder Microscopical characters with diagram/ photographs and description

Micro-chemical study

Micro-chemical tests	Name of test	Observation	Inference
Lignin			
Starch grain			
Calcium crystals			
Oil (Volatile/non-volatile)			
Major secondary metabolite			

Powder Characteristics:

Color	
Odor	
Taste	
Powder Microscopical features	

Chemical study:

Chemical class	Name of test	Observation	Inference
Essential oil test			
Tannins test			

Result: From morphological, microscopical, micro chemical, powder and chemical study it is confirmed that given crude drug is ....

Questions

1. What is vittae?
2. Which is confirmatory chemical test for fennel major chemical constituent?
3. Which is major chemical constituent of fennel and what are it's commercial uses?
4. Draw neat labeled diagram of TS of fennel.
5. Draw neat labeled diagram of fennel powder characters.
6. Which type of calcium crystals is present in fennel fruit?

Aim 06: To study morphological, histological, powder microscopical and chemical characteristics of crude drug- *Senna*

Requirements: Microscopes, slides, test tubes, all pharmacognosy reagents and chemical test reagents.

Theory

Senna: Senna (Senna leaf, Indian senna, Senna folium, Sennai-ki-patti, Tinnevelly senna, Cassia senna) is dried leaflets of *Cassia angustifolia* Vahl of Family Leguminosae. It contains antraquinone glycosides Sennoside A, B, C & D, rhein 8-glucoside, rhein 8-diglucoside, aloe-emodin, 8-glucoside, anthrone, diglucoside, rhein, kaempferol, aloe-emodine and isorhamnetin. It is used as purgative.

Experimental Morphological study:

Color	
Odor	
Taste	
Size	
Shape	
Extra feature	

Microscopical study

Transverse section shows upper and lower epidermises composed of thin sinuous walled polygonal cells with cuticular striations, numerous paracytic stomata, unicellular covering conical, thick warted walled trichomes which are curved at the base, single layer of palisade cells below upper and lower epidermis, palisade and spongy parenchymatous sheath surrounding the groups of fibers having prismatic calcium oxalate crystals, thick-walled and lignified fibers. Many of the epidermal cells contain mucilage.

Labeled diagram of Morphological characters of crude drug

Labeled photograph of TS of crude drug

Powder Microscopical characters with diagram/ photographs and description

Micro-chemical study

Micro-chemical tests	Name of test	Observation	Inference
Lignin			
Starch grain			
Calcium crystals			
Oil (Volatile/non-volatile)			
Major secondary metabolite			

Powder Characteristics:

Color	
Odor	
Taste	
Powder Microscopical features	

Chemical study:

Chemical class	Name of test	Observation	Inference
Anthraquinone glycoside test			
Saponin test			

Result: From morphological, microscopical, micro chemical, powder and chemical study it is confirmed that given crude drug is ……………

Questions

1. Which is confirmatory chemical test for senna major chemical constituent?
2. Which is major chemical constituent of senna and what are it's commercial uses?
3. Draw neat labeled diagram of TS of senna.
4. Draw neat labeled diagram of senna powder characters.
5. Which type of stomata, trichome and calcium crystals are present in senna leaves?

Aim 07: To study morphological, histological, powder microscopical and chemical characteristics of crude drug- *Cassia/Cinnamon*

Requirements: Microscopes, slides, test tubes, all pharmacognosy reagents and chemical test reagents.

Theory

Cassia: Cassia (cassia cinnamon, Cassia bark, Chinese cinnamon) is dried stem bark of the plant *Cinnamomum cassia*, Lauraceae. It contains monoterpenoids named as cinnamic aldehyde, eugenol and cinnamyl acetate. It is used as carminative, stimulant, flavourant, and as a spice.

Experimental Morphological study:

Color	
Odor	
Taste	
Size	
Shape	
Extra feature	

Microscopical study

Transverse section shows polygonal, thick walled cells of cork with conspicuous granular, reddish-brown contents, the medullary ray cell containing numerous small acicular crystals of calcium oxalate, abundant sclereids, thick walled lignified fibers with inconspicuous pits, parenchymatous tissues consisting scattered spherical to ovoid, simple or compound starch grains and volatile oil cells.

Labeled diagram of Morphological characters of crude drug

Labeled photograph of TS of crude drug

Powder Microscopical characters with diagram/ photographs and description

Micro-chemical study

Micro-chemical tests	Name of test	Observation	Inference
Lignin			
Starch grain			
Calcium crystals			
Oil (Volatile/non-volatile)			
Major secondary metabolite			

Powder Characteristics:

Color	
Odor	
Taste	
Microscopical features	

Chemical study:

Chemical class	Name of test	Observation	Inference
Essential oil test			
Tannins test			

Result: From morphological, microscopical, micro chemical, powder and chemical study it is confirmed that given crude drug is ....

Questions

1. What are morphological, microscopical and chemical differences between cassia and cinnamon?

2. Which is confirmatory chemical test for cinnamon major chemical constituent?

3. Which is major chemical constituent of cassia and what are it's commercial uses?

4. Draw neat labeled diagram of TS of cassia bark.

5. Draw neat labeled diagram of cassia bark powder characters.

6. Which type of xylem and calcium crystals is present in cassia bark?

Aim 08: To isolate caffeine from tea dust and confirm by chemical tests

Requirements: Water, chloroform, separating funnel, beakers

Caffeine

Theory

Most of the people start their morning with the cup of tea which contains the alkaloid caffeine. Caffeine is a naturally occurring substance found in the leaves, seeds or fruits of over 63 plants species worldwide and is part of a group of compounds known as methylxanthines. The most commonly known sources of caffeine are coffee, cocoa beans, cola nuts and tea leaves. Caffeine is a bitter, white crystalline xanthine alkaloid that is a psychoactive stimulant drug. Caffeine is pseudo alkaloid as it is not biosynthesised from amino acids but gives all alkaloid identification tests positive. Extraction of caffeine is a simple decoction process where caffeine is firstly dissolved in water and separated by chloroform. Tea leaves contain 1 - 4% of caffeine while coffee seeds contains 1 - 2% of caffeine.

Procedure

Weigh 50 gm of tea leaves and transfer to 200 ml distilled water. Boil the water for 30 minutes with occasional stirring. After complete decoction is effected, allow to cool and filter the solution. Take filtrate in a separating funnel and to it add 25 ml chloroform. Shake slowly so that total caffeine will be transferred to chloroform. Separate chloroform layer. Evaporate chloroform over water bath. White caffeine crystals will collect at the bottom.

Result:

Color	
Consistency	
Yield	
Confirmatory chemical test Murexide test	

Questions

1. Draw chemical structure and IUPAC name of Caffeine.
2. What are different methods of Caffeine isolation?
3. How Caffeine is different from theophylline and theobromine?
4. Which is confirmatory chemical test for caffeine?
5. How many plants contain caffeine?
6. What are commercial applications of isolated caffeine compared to tea leaves powder?

Aim 09: To extract eucalyptus oil by Clevenger apparatus (Hydro distillation)

Requirements: Water, fresh eucalyptus leaves, Clevenger apparatus, test tubes

Theory

Generally, volatile oils are extracted from plant material by four methods, i.e., distillation, steam distillation, expression, and Enfleurage. Most oils are distilled in a single process within 2-5 hours, while a few of them require a second step to purify them through fractional distillation. Clevenger apparatus is specially designed apparatus for extraction of volatile oil by water distillation method. Generally it is used to estimate water content.

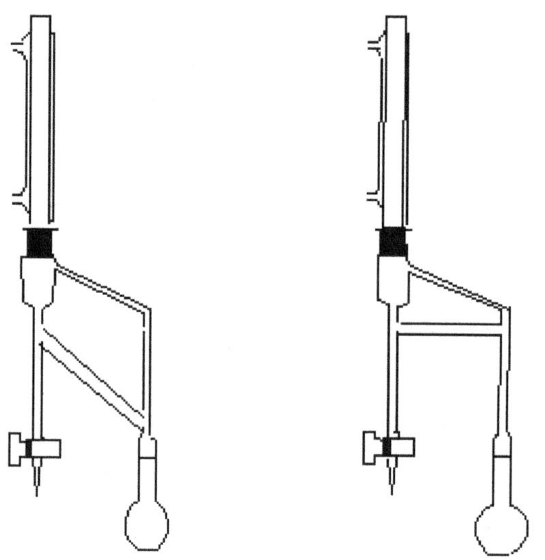

Clevenger apparatus for volatile oil determination

Eucalyptus oil is readily steam distilled from the leaves and can be used for cleaning and as an industrial solvent, as an antiseptic, for deodorising, and in very small quantities in food supplements, especially sweets, cough drops, toothpaste and decongestants. It has insect repellent properties, and is an active ingredient in some commercial mosquito repellents. *Eucalyptus globules,* family Myrtaceae is the principal source of eucalyptus oil worldwide.

Procedure

Weigh 100 gm of fresh eucalyptus leaves and transfer to 250 ml distilled water in RBF of Clevenger apparatus. Boil the water for 60 minutes and collect oil from graduated tube. Calculate percentage yield.

Result:

Color	
Consistency	
Yield	
Confirmatory chemical test	

Questions

1. What are different methods of volatile oil isolation?
2. Which is major chemical constituent of Eucalyptus oil?
3. What are two different types of Clevenger apparatus?
4. What is commercial use of eucalyptus oil?
5. Which is confirmatory chemical test for eucalyptus oil?
6. What is approximate price of 1ml of eucalyptus oil?
7. What is principle of eucalyptus oil extraction?

Aim 10: To perform TLC of extracted eucalyptus oil

Requirements: Silica Gel G, Water, TLC plates, TLC chamber, Chloroform, Methanol, extracted eucalyptus oil.

Theory

Chromatography literally means "colour writing". It is a technique of separating a mixture of components, by distributing between two phases that is stationary phase and mobile phase depending on the relative affinities of the components for the two phases. Stationary phase may be solid, gel or liquid supported by solid. Mobile phase may be liquid or gas. Planar chromatography (Paper, TLC, HPTLC) and Column chromatography (Open column, GC, HPLC, VLC, IEC, Gel chromatography, Flash chromatography) are the two main types of chromatographic techniques.

Eucalyptus (*Eucalyptus globules,* Myrtaceae) leaves oil contains 1.25-0.75% volatile oil composing chiefly cineol also called as eucalyptol (more than 65 %), Pinene, Phellandrene, Citronellal, and limonene.

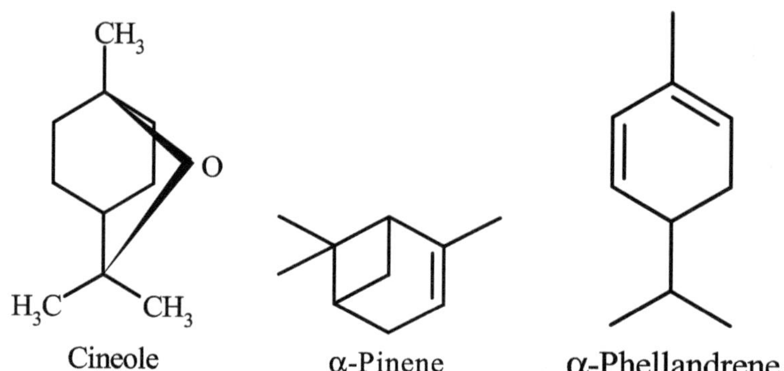

Procedure

Stationary Phase: Silica Gel G   Mobile Phase: Toluene: Ethyl acetate (7:3)

Detection: Anisladehyde-Sulfuric acid spray And heat at 110°C for 5 min

➢ Dissolve 1 ml of sample in 5 ml of methanol.

➢ Prepare adsorbent slurry in 1: 3 proportions. (E.g. Silica Gel G: Water-1:3)

➢ Prepare TLC plates by this slurry by Pouring technique.

➢ Prepare mobile phase as per composition and allow saturating TLC chamber.

➢ Apply sample by cotton wig in the form of band. Develop chromatogram by ascending technique. Blue, violet, pink colored spots of volatile terpenes and terpenoids will be observed.

Observation:

Components	Color before derivitisation	Rf value	Color after derivitisation	Rf value	Inference
Cineole					
Pinene					
Phellandrene					
Piperitone					
Citronelllal					

Result: Eucalyptus Oil is separated by TLC and Rf values with color of components at are confirmed as by comparison with authentic book.

Questions

1. What is Rf value?
2. What is meaning of "G" in name stationary phase Silica Gel G?
3. Which is commonly used mobile phase and stationary phase for TLC of eucalyptus oil?
4. How many chemical constituents are present in eucalyptus oil?
5. Which is major chemical constituent of Eucalyptus oil?
6. What is principle of TLC?

Aim 11: To isolate Sennosides from Senna and confirm by chemical tests

Requirements: Senna leaf powder, benzene, electric shaker, methanol, Hydrochloric acid, Calcium chloride, and ammonia

Theory

Sennosides are anthraquinone glycosides obtained from dried leaflets of *Cassia auriculata* (Alexandrian senna) and *Cassia angustifolia* (Tinnevelly senna), of the family Leguminaceae. Sennoside A, B, C and D are isomers of each other. These are useful as potent laxatives.

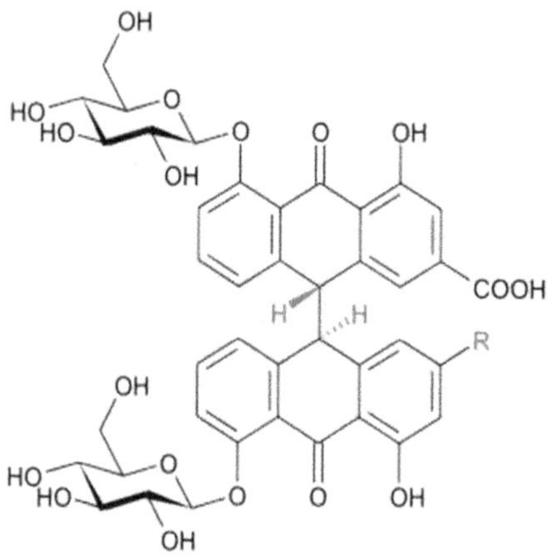

Sennosid A: R = COOH
Sennosid C: R = CH₂OH

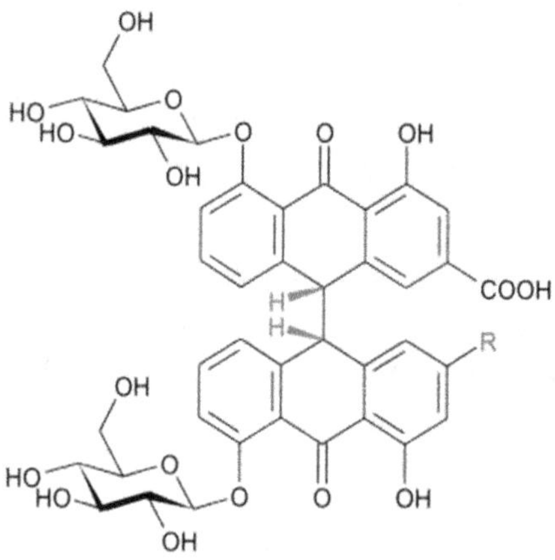

Sennosid B: R = COOH
Sennosid D: R = CH₂OH

Procedure

1. Take the 50 gm weighed quantity of coarse powder of *Senna* leaves. Extract the powder with 100 ml benzene on an electric shaker for 1hours to remove lipids and pigments.

2. Filter and discard the benzene extract. Extract the dried marc with 150 ml of 70 % methanol for 2 hours. Filter and again extract the marc with fresh methanol for 1hours.

3. Filter and combine both extracts and concentrate to 1/4ᵗʰ of its original volume.

4. Add sufficient quantity of HCl to obtain a pH of 3.2 and keep aside for 30 min at 5°C.

5. Filter and add sufficient quantity of alcoholic anhydrous calcium chloride (2 gm in 25 ml denatured spirit) with continuous stirring. Add ammonia to bring the pH to 8 and keep aside for 2 hours.

6. Filter and collect the precipitates of calcium sennoside. Dry the precipitate in a desiccators and calculate the percentage yield.

Result

Color	
Consistency	
Yield	
Confirmatory chemical test	

Questions

1. Draw chemical structure and IUPAC name of Sennosides.
2. Which is confirmatory chemical test for sennosides?
3. Which is key step in sennoside extraction?
4. Why to extract sennosides in calcium salt form?
5. What is commercial use of sennoside compared to seena leaves powder?

Aim 12: To isolate Disogenin from Dioscorea/Methi seeds and confirm by chemical tests

Requirement: Methi seeds powder, methanol, HCl,

Theory

Diosgenin, a phytosteroid sapogenin, is the product of hydrolysis by acids, strong bases, or enzymes of saponins, extracted from the tubers of Dioscorea wild yam, such as the Kokoro. The sugar-free (aglycone) product of such hydrolysis, diosgenin is used for the commercial synthesis of cortisone, pregnenolone, progesterone, and other steroid product

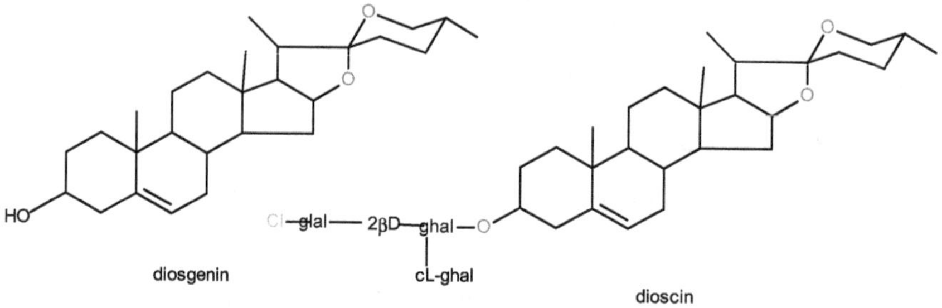

Procedure

➢ Take 50 gm of coarse powder of methi seeds

➢ Reflux powder with 150 ml 2N HCl for 30 min.

➢ Extract marc with, 200ml methanol for 1 hr by reflux.

➢ Filter and distill off the solvent to get $1/4^{th}$ volume of original mass.

➢ Keep this semisolid mass in refrigerator for 30 min to from crystals of Disogenin.

Result:

Color	
Consistency	
Yield	
Confirmatory chemical test	

Questions

1. Draw chemical structure and IUPAC name of Disogenin.
2. How many plants contain diosgenin? List them.
3. What is commercial use of disogenin?
4. Which is key step in diosgenin extraction?
5. How to separate disogenin from other chemical constituents present in methi seeds?
6. Which is confirmatory chemical test for disogenin?
7. What can be approximate percentage of purity of isolated diosgenin?

Aim 13: To isolate Atropine from Belladonna/Datura and confirm by chemical tests

Requirement: Belladonna/Datura leaf powder, alcohol, HCl, ether, acetic acid, sodium carbonate, chloroform, ammonia, potassium carbonate, sulphuric acid

Theory:

Atropine is a tropane alkaloid obtained from *Atropa belladonna* (deadly night shade), *Datura stramonium* (thorn apple), and *Hyoscyamus niger* (henbane) of family *Solanaceae*. It is found in many members of the *Solanaceae* family. It is a diastereomeric mixture of d-hyoscyamine and l-hyoscyamine with most of its physiological effects due to l-hyoscyamine. Its pharmacological effects are due to binding to muscarinic acetylcholine receptors. It is an antimuscarinic and anticholinergic agent. It is mydriatic and antidote in opium poisoning. Atropine is contraindicated in patients pre-disposed to narrow angle glaucoma.

IUPAC name: [(1R,5S)-8-methyl-8-azabicyclo[3.2.1]octan-3-yl] 3-hydroxy-2-phenylpropanoate

Extraction Methods

Weigh 50 gm of crude drug powder and moist with aqueous solution of Sodium carbonate and then extract with benzene or ether. Extract free bases using acetic acid acidified water.

Separate acid layer and add ether to remove interfering substances. Add sodium carbonate to precipitate alkaloids. Separate precipitate and dissolve in ether, acetone. Filter and evaporate filtrate to dry crystalline mass. Dissolve this dry crystalline mass in alcohol and add sodium hydroxide to get optically inactive hyoscyamine due to complete recemisation of atropine

Result:

Color	
Consistency	
Yield	
Confirmatory chemical test	

Questions

1. Draw chemical structure and IUPAC name of Atropine.
2. Which is confirmatory chemical test for Atropine?
3. Which is key step in Atropine extraction?
4. What is commercial use of Atropine and list marketed products with their company details.

Aim 14: To analyse crude drugs (Aloe, Benzoin, Myrrh, Asafoetida, and Colophony) by chemical tests

Requirements: Test tubes, all Pharmacognosy reagents and chemical test reagents.

Procedure:

Drug	Test	Observation	Inference
Aloe Dried latex of leaves of Aloe species like *Aloe vera, Aloe barbadensis* Miller (or Curacao Aloe); *Aloe ferox* Miller (or Cape Aloe); *Aloe perryi* Baker (or Socotrine Aloe); *Aloe africana* Miller and *Aloe spicata* Baker (or Cape Aloe) of family Asphodelaceae	**Cupraloin (Klunge's isobarbaloin) test** Take little quantity of aqueous solution of aloes, add a few drops of saturated $CuSO_4$, a few mg of sodium chloride, and 5 ml of 90% alcohol.	Curacao aloes: wine red color lasts for many hours Cape aloes: faint color immediately changes to yellow Socotrine aloes: no color Zanziber aloes: no color	
	Nitric acid test Add nitric acid to aqueous solution of aloes.	Curacao aloes: dark red color. Cape aloes: brown color changes to green. Socotrine aloes: light brown color changes to yellow. Zangzber aloes: yellowish brown color.	
	Nitrous acid test Add sodium nitrite and acetic acid to aqueous solution of aloes.	Curacao aloes: dark pink color. Cape aloes: faint pink color. Socotrine and	

Contd...

		zanziber aloes: less change in color.	
	Borax bead test Heat 5 ml solution with 0.2 gm borax; add few drops of this solution in a test tube filled with water.	Green fluorescence is produced	
Benzoin Balsamic resins obtained from *Styrax benzoin* or *Styrax paralloneurus* (Sumatra benzoin) and *Styrax tonkinesis* (Siam benzoin) of family styraceae	Heat 0.5 g slowly in a dry test tube.	It melts and evolves irritating whitish fumes which condense to form a whitish crystalline sublimate in the upper part of the tube.	
	Heat gently 1 g of powder with 5 ml of potassium permanganate solution in a test tube.	Distinct odor of benzaldehyde	
	Titrate 0.1 g of powder with 5 ml of alcohol (95%), filter and to the filtrate, add 0.5 ml of 5% w/v alcoholic ferric chloride solution.	No bright green color (distinction from siam benzoin).	
	Dissolve 0.2 g of benzoin in ether. Add 1 ml of the ethereal solution in a porcelain dish containing 1 ml of concentrated sulphuric acid.	Reddish-brown color	
Myrrh Olio –gum resin obtained from *Commiphoramolmol*of family Burseraceae	Titrate powder with water.	Yellowish emulsion	
	Treat dry ethereal extract with bromine vapors.	Red coloration	
	Treat dry ethereal extract with nitric acid solution.	Purple coloration	
Asafoetida Exudates from *Ferulafoetida,*	Titrate asafoetida with water with addition of alkali.	Yellowish-orange emulsion turning to greenish-yellow	

Contd...

Family- Umbelliferae	Heated with dilute sulphuric acid.	Reddish- brown coloration	
	Take 10 ml of alcoholic extract; and add few drops of phloroglucinol and concentrated hydrochloric acid.	Pink color	
	Boil with dilute hydrochloric acid and filter into ammonia.	Blue fluorescence due to free umbelliferone	
Colophony Residue remained after distillation of crude oleo resin of *Pinus palustris and other Pinus species* of family Pinaceae.	Dissolve 0.1 g powder in 10 ml of acetic anhydride with mild heat, cool, and add a drop of concentrated sulphuric acid.	Purple-red color changes to violet due to reaction of abietic acid	
	Dissolve 0.1 g powder in petroleum ether and filter. To filtrate, add 2-3 ml dilute copper acetate solution.	Ether portion shows emerald green color due to reaction of abietic acid	
	Expose alcoholic solution of colophony to litmus paper.	Color changes due to acidic reaction	

Results: Aloe, Benzoin, Myrrh, Asafoetida, Colophony drugs are analysed and confirmed by chemical tests.

Questions

1. What are chemicals tests to identify aloe?

2. What are chemicals tests to identify Benzoin?

3. What are chemicals tests to Myrrh?

4. What are chemicals tests to identify Asafoetida?

5. What are chemicals tests to identify Colophony?

6. What is biological source of Aloe, Benzoin, Myrrh, Asafoetida and Colophony?

7. What is chemical composition of Aloe, Benzoin, Myrrh, Asafoetida and Colophony?

8. Which is major chemical constituent of Aloe, Benzoin, Myrrh, Asafoetida and Colophony?

9. Why colophony shows acidic reaction to litmus paper?

10. Which chemical constituent is responsible for Blue fluorescence of acidic extract of Asafoedita in ammonia?

11. Which colors are produced by different aloes when treated with nitrous acid?

12. Which colors are produced by different aloes when treated with Nitric acid?

13. Which chemical constituent of Aloe is identified by Klunge's tests?

Aim 15: To perform separation of sugars by Paper chromatography

Requirements: Whatman paper, chromatographic chamber, water-saturated phenol, 1% ammonia, n-butanol, acetic acid, water, isopropanol, pyridine, acetic acid, glucose, fructose, maltose, lactose etc.

Procedure:

Stationary Phase	Paper : Usually Whatman No. 1 filter paper
Mobile Phase	A. n-butanol-pyridine-water (60: 40: 30) B. n-butanol-acetone-water (20: 70: 10) C. Saturated alcoholic solution of picric acid
Spray reagents	A. Resorcinol reagent: Mix 1% ethanolic solution of resorcinol and 0.2N HCl (1:1 v/v).Spray the dried chromatograms and visualize spots by heating at 90C. B. Spray with 5% alcoholic sodium hydroxide. Dry under sunlight

1. Place sufficient solvent into the bottom of the tank. Cover the led and allow the tank to be saturated with the solvent.
2. Take a sheet of whattman 1 chromatography paper (about 9 x 10 cm) and place it on a piece of clean paper on a bench.
3. Draw a fine line with a pencil along the width of the paper and about 1.5cm from the lower edge.
4. Along this line place four equality spaced (about 2cm apart) small circles with a pencil.
5. Label the paper at the top with the name of each of the sugars and label the last unknown.
6. Use a fine capillary or tooth pick to place the drops of the solutions of the sugars, glucose, fructose, maltose, lactose and the mixture.
7. After spotting, dry the paper with hot air dryer for one minute, repeat this step again.
8. Place the spotted paper in the chromatographic tank and make the development by using the ascending technique.
9. Close the tank with lid, allow the solvent to flow for about 30-45 minutes.
10. Remove the paper and immediately mark the position of the solvent front with a pencil.
11. After the chromatogram has dried, spray the paper with the locating reagent.
12. Circle the position of each spot with pencil. Calculate the Rf value for each spot and also for the spots the mixture contained.

Results:

Sample	Color before derivitisation	Rf value	Color after derivitisation	Rf value	Inference
Glucose					
Fructose					
Maltose					
Dextrose					
Mixture					

Questions

1. What advantage and disadavantages are of paper chromatography compared to TLC?
2. What are techniques used to separate mixtures by Paper chromatography?
3. What are the factors affects paper chromatography?
4. Which commercial chromatography papers are available in market?
5. What is principle of paper chromatography?

Further Reading

1. A.N.M. Alamgir. Therapeutic Use of Medicinal Plants and Their Extracts: Volume 1

2. Alice Kurian, M. Asha Sankar Medicinal Plants. New India Publishing Agency. 2007

3. Anjoo Kamboj, Moronkola Dorcas Olufunke Practical Pharmacognosy. Scitus Academics LLC. Thomas Edward Wallis. Practical Pharmacognosy. Churchill. 2018

4. Ashutosh Kar. Pharmacognosy And Pharmacobiotechnology. New Age International (P) Limited. 2003

5. Biren Shah, Avinash Seth. Textbook of Pharmacognosy and Phytochemistry. Elsevier Health Sciences. 2014

6. C. S. Shah, J. S. Qadry. A Textbook of Pharmacognosy. Messrs B.S. Shah 1971

7. C.K. Kokate, Purohit, Gokhlae. Text book of Pharmacognosy, 37th Edition, Nirali Prakashan, Pune. 2007

8. Christophe Wiart, Ashok Kumar. Practical Handbook of Pharmacognosy-Preliminary Techniques of Identification of Crude Drugs of Plant Origin. 2000.

9. Deore SL, Khadabadi SS, Baviskar BA. Pharmacognosy and Phytochemistry: A Comprehensive Approach. Second Edition. BSP Publicatons, Hyderabad. 2017.

10. GS Kumar. KN Jayaveera. A Textbook of Pharmacognosy and Phytochemistry. S CHAND & Company Limited. India. 2014.

11. Gunnar Samuelsson. Drugs of Natural Origin-A Textbook of Pharmacognosy. Apotekarsocieteten. 1999.

12. H. Ansari. Essentials of Pharmacognosy. Second edition, Birla publications, New Delhi, 2007

13. James Bobbers, Marilyn KS, VE Tylor. Pharmacognosy & Pharmacobiotechnology. Williams & Wilkins. 1996.

14. Jean Bruneton. Pharmacognosy, Phytochemistry, Medicinal Plants. Technique & Documentation. 1999.

15. Joshi Saroja, Vidhu Aeri. Practical Pharmacognosy. Frank Brothers. 2009.

16. Khadabadi SS, Deore SL, Baviskar BA. Experimental Phytopharmacognosy A Comprehensive Guide. Nirali Prakashan, Pune. 2019

17. K. Mangathayaru. Pharmacognosy: An Indian perspective. Pearson Education India. 2013

18. K. R. Khandelwal. Practical Pharmacognosy. Nirali Prakashan, Pune. 2008

19. Kaliya.A. Text Book of Industrial Pharmacognosy. CBS Publishers & Distributors, Delhi. 2009

20. Kendall Jefferson. Pharmacognosy and Phytotherapy. Foster Academics.2019

21. Kokate CK. Practical Pharmacognosy. Vallabh Prakashan, Delhi. 2005

22. Luqi Huang. Molecular Pharmacognosy. Springer Netherlands. 2012

23. M A Iyengar, S G K Nayak. Pharmacognosy Lab Manual. PharmaMed Press. 2019

24. M A Iyengar. Pharmacognosy of Powdered Crude Drugs. PharmaMed Press. 2017

25. M A Iyengar. Study of Crude Drugs. PharmaMed Press. 2016

26. Michael Heinrich, Elizabeth M. Williamson, Joanne Barnes, Simon Gibbons, Jose Prieto-Garcia Fundamentals of Pharmacognosy and Phytotherapy E-Book. Elsevier Health Sciences. 2017

27. Michael Heinrich, Joanne Barnes, Simon Gibbons. Fundamentals of Pharmacognosy and Phytotherapy. Churchill Livingstone/Elsevier. 2012

28. Mohammad Ali. Pharmacognosy and Phytochemistry, CBS Publishers & Distribution, New Delhi.

29. N P S Sengar, Ashwini Singh, Ritesh Agrawal. A Textbook of Pharmacognosy. PharmaMed Press. 2018

30. Nilambari S. Gurav, Shailendra S. Gurav Indian Herbal Drug Microscopy. Springer New York. 2013

31. Rangari VD. Pharmacognosy& Phytochemistry. Career Publication, Nashik. 2008

32. S. S. Agarwal, M. Paridhavi. Herbal Drug Technology. Universities Press. 2012

33. S. S. Handa. Pharmacognosy. Vallabh Prakashan, New Delhi. 1989

34. Saikat Sen, Raja Chakraborty Herbal Medicine in India-Indigenous Knowledge, Practice, Innovation and Its Value. Springer Singapore. 2019

35. Simone Badal Mccreath. Rupika Delgoda. Pharmacognosy-Fundamentals, Applications and Strategies. Elsevier Science. 2017

36. Steven E. Ruzin. Plant Microtechnique and Microscopy.1999

37. T. C. Denston. A Textbook of Pharmacognosy. Read Books. 2012

38. Vidhu Aeri, D.B. Anantha Narayana, Dharya Singh. Powdered Crude Drug Microscopy of Leaves and Barks. Elsevier Science. 2019

39. W.C.Evans, Trease and Evans Pharmacognosy, 16th edition, W.B. Sounders & Co., London, 2009.

40. The British Pharmacopeia. London: Medicines and Healthcare Products Regulatory Agency; 1993.

41. Indian Pharmacopeia 2018, Ghaziabad: Indian Pharmacopeia Commission; 2018.

42. United States Pharmacopoeia and National Formulary, USP 25 NF 19/National Formulary 20, Rockville, MD, U. S. Pharmacopoeial Convention, Inc. 2002.

43. Ayurvedic pharmacopoeia of India Part-I vol.I, 2001.